A History of Anthropology as a Holistic Science

A History of Anthropology as a Holistic Science

By Glynn Custred

LEXINGTON BOOKS
Lanham • Boulder • New York • London

Published by Lexington Books
An imprint of The Rowman & Littlefield Publishing Group, Inc.
4501 Forbes Boulevard, Suite 200, Lanham, Maryland 20706
www.rowman.com

Unit A, Whitacre Mews, 26-34 Stannary Street, London SE11 4AB

British Library Cataloguing in Publication Information Available

Library of Congress Cataloging-in-Publication Data

Names: Custred, Glynn (Professor Emeritus of Anthropology)
Title: A history of anthropology as a holistic science / Glynn Custred.
Description: Lanham : Lexington Books, 2016. | Includes bibliographical references and index.
Identifiers: LCCN 2016005180 (print) | LCCN 2016010234 (ebook) | ISBN 9781498507639 (cloth : alk. paper) | ISBN 9781498507646 (Electronic)
Subjects: LCSH: Philosophical anthropology. | Anthropology.
Classification: LCC BD450 .C867 2016 (print) | LCC BD450 (ebook) | DDC 301.01--dc23
LC record available at http://lccn.loc.gov/2016005180

∞™ The paper used in this publication meets the minimum requirements of American National Standard for Information Sciences Permanence of Paper for Printed Library Materials, ANSI/NISO Z39.48-1992.

Printed in the United States of America

I would like to thank colleagues who have read portions of this book and have given me useful comments on how to improve it: Ben Blount, Alan Almquist, John Ellis, Peter Wood, Roland Alum, and Andrei Simic. I also wish to express my gratitude to those who introduced me to the four fields of our discipline and how to conduct them: Paul Doughty, Carl Voegelin, David Bidney, Alan Merriam, Harold Driver, George Neuman, and James Kellar, Indiana University 1967–1973, as well as to my mentors in Andean anthropology: John V. Murra and John H. Rowe.

Table of Contents

Introduction

From the very beginning, the four fields of study that constitute contemporary American anthropology were perceived as scientific enterprises. Toward the end of the nineteenth century they merged into the discipline that has since been described as "the general theory of man" (Boas), "the whole science of man" (Lowie), "the science of man at large" (Malinowski), and "the science of man and his works" (Kroeber). Anthropology's scope is thus broad, ranging from the physical and the social sciences to the humanities. Its stated purpose is to understand the human species as a whole: its biological nature and evolution; its subsequent passage through time; its social and cultural organization in its wide range of variation; and the study of the most human feature of all, language and its relationship to culture and society. For that reason, anthropology has been characterized as a holistic discipline, one which seeks knowledge of the whole through knowledge of its various parts.

Eventually a centrifugal tendency set in, as the number of anthropologists grew and as the discipline reached out and penetrated ever deeper into different areas of the human experience. This tendency eventually weakened the appeal of the original holistic concept, with some specialists gravitating to neighboring disciplines, leading some to call for abandonment of the traditional four field approach. In the 1960s and 1970s a new element entered the picture which exacerbated the centrifugal tendencies already in play, a demand that anthropology abandon its primary mission as a scientific enterprise and become, at least in part, an instrument of social change, thus prompting some in the profession to openly defend anthropology as a science (see Kuznar 1972, second edition 2008).

By the 1980s both fragmentation and demands for redefinition had led some anthropologists to perceive a crisis in the discipline (Hoebel, et al 1982). The split between the activists and those who wished to retain anthropology's original mission contributed in some universities to the separation of different branches of anthropology into different departments. This split led to the formation of a new association which was registered in 2004 as a non-profit organization in California, and becoming a section of the AAA in 2008 under the name Society of Anthropological Sciences (SAS).

This increasing split in the profession eventually appeared on the highest organizational level when, in 2010, the executive committee of the

American Anthropological Association dropped the word "science" from its long-term mission statement (and in several other places as well), a move that attracted the attention not only of anthropologists but also of many outside the field. Peter Peregrine, then president of the SAS told the *New York Times* that the change in the mission statement "just blew the top off" a conflict that had been building for a long time. "Even if the board goes back to the old wording," he said, "the cat's out of the bag and running around clawing up the furniture" (Wade 2010).

The purpose of this book is to look back at the history of the discipline, at the way it has developed over the last two centuries and at the contributions it has made, and to argue that despite fragmentation and despite the drive for redefinition, the scientific and the holistic approach should be maintained and further developed. For this purpose a historical review such as the one presented here is necessary since, as Herbert Lewis points out, the current redefinition and redirection of the discipline, "conceived in anger and produced under the influence of fashionable ideology," has created misrepresentations which have had a "seriously damaging effect on the whole 'project' of anthropology" by turning away generations from its "hard-won learning," leaving them "unaware if not scornful of data, analyses, ideas and approaches of earlier generations" (Lewis 2014: xiii–xiv).

We start by looking at the time when it all started, at what has been called the Second Scientific Revolution, a time of intense scientific activity, when the rationalism of the eighteenth century and the Romanticism of the first half of the nineteenth century merged to create the milieu in which anthropology first appeared. In the two subsequent chapters we see how science caught on and how it developed as both a cultural category and a social process, creating the context in which anthropology emerged. We linger awhile on that period not only because it is a fascinating story, but also because it illustrates how deeply the origins of anthropology were rooted in a scientific and in a holistic vision, as well as revealing a duality of those times that has characterized anthropology ever since, one that will strike a responsive chord in many who have heard the call of earlier times and of distant places and of remote people (and of people closer at hand) and who feel the desire to understand them and to make that understanding known to others, and who have thus chosen to follow those pursuits as their vocation.

We then see how the discipline developed in four countries, Britain, Germany, France, and the United States, each representing different variations of a common and interconnected intellectual tradition, but which despite the differences has resulted in a single discipline. We turn next to the question, what is science? We begin with the meaning of the word "science" and its change over time to show how the concept (which the word labels) has developed in response to changing social, cultural and intellectual factors. In that way we reveal—as historian of science David

Lindberg points out—that the meaning of the word "science," like the meaning of other words, does not rest on "set rules grounded in the universe," but is instead "a set of conventions adopted by a group of people" (1992: 2). The current meaning of science at this point in time and in English-speaking countries is thus only one of other possibilities.

We then examine one of those other possibilities, namely the German word for science and the concept it labels. The meaning of the English word has narrowed over time. The same process was at work in Germany until the middle of the nineteenth century when the narrowing process was stopped and the original broad meaning was restored and refined, a meaning which comfortably accommodates the holistic nature of anthropology. In fact, the development of American anthropology was in a significant way a product of that tradition, which is the topic of chapter 7.

Another question which arises is whether it is advisable to retain a holistic approach in a discipline which spans such a broad range of subject matter. In chapter 6 we address this question by pointing out that divergence in a growing field of knowledge is sometimes followed by the equally natural tendency of convergence in various forms, and even in some cases in the creation of new sciences. In this regard we look at the emergence of the new field of cognitive science created by the convergence of different branches of science on a single topic, and at the interdisciplinary nature of the study of language which has developed by the same process of convergence but without creating a separate branch of science. In this way we show that a holistic approach is not only feasible, but that it is also a natural tendency in the advancement of the scientific enterprise.

We then move to each of the four sub-disciplines of anthropology in order to show how they deal with the specific sections of the human species they study, and how those sub-disciplines form parts of a broader whole. In chapter 15 we look at anthropology as it is applied to the study of a distinctive culture area, in order to show the interrelationship of the sub-disciplines and how anthropological inquiry relates to neighboring branches of science. The culture area we examine here is the Central Andean highlands, one of the "cradles of Civilization," with its distinctive patterns of plant, animal, and human adaptation to an ecologically varied high mountain environment—with its long and archaeologically retrievable culture history and its shorter but rich historical record, as well as an ethnographic present which reaches back into early times through discernible cultural continuities, and which, like all societies and cultures today, faces both the challenges and the opportunities of a changing world. Within that context, and in order to further illustrate the ethnographic method, we look briefly at the author's ethnographic experience in a high altitude Quechua-speaking community in Southern Peru.

In the epilogue we sum up by arguing that the original holistic concept of anthropology is both feasible and advantageous, and that activism, separate from the central scientific mission of the discipline, can and indeed always has been practiced by anthropologists, but that the two must be kept separate for the further development of a discipline which has already delivered so much and which promises even more in the future.

ONE

The Second Scientific Revolution

Science has been compared to a relay race where one generation hands off to its successors what it has achieved in its lifetime. The receiving generation then uses those achievements as a platform from which further advances are launched. The first stage in the relay goes back to the sixteenth and seventeenth centuries in what has been called the Scientific Revolution, starting when Nicholas Copernicus (1473–1543) presented a new model of the universe in which the earth rotated around the sun and not the other way around. This served as the platform for Johann Kepler's (1571–1630) theory, bolstered by the careful and systematic observations of Tycho Brahe (1546–1601), a model which was the basis for Galileo's (1564–1642) theories. Those were the pillars upon which Newton based his theory of gravitation.

That was also the time when William Harvey (1578–1657) described in detail the circulation and the properties of blood. Robert Boyle (1627–1691) made major discoveries in chemistry and Sir Isaac Newton, who was lionized for his achievement and buried in Westminster Abbey, made a major breakthrough in astronomy and physics with his theory of gravitation and celestial mechanics.

Also important was the work of Francis Bacon (1561–1626) who advanced the scientific method, John Locke (1627–1691) who formulated the epistemological foundation for the science of his time, and René Descartes (1596–1650) who laid the foundation for later developments in mathematics. These events define the Scientific Revolution which provided the basis for an ever growing, and an ever more successful enterprise which was to reshape Western Civilization and spread throughout the world.

The following century was known as the Enlightenment or the Age of Reason, a time when rationality and the idea of progress comprised the spirit of the time. No longer was the quest for truth to be sought by means of deductive logic applied to a body of authoritative writing, as had been the case in the Middle Ages, when the works of ancient Greek natural philosophy and Christian theology were regarded as the ultimate authority. Instead, inductive logic was to be applied to empirical observation and experimentation in search of still undiscovered laws of universal validity. For example, the motto of the Royal Society in London was *nullius in Verba,* meaning in essence, don't simply take somebody else's word for it; go out and test it yourself.

At the end of the eighteenth and in the early nineteenth centuries the baton passed to a new generation of scientists, resulting in a new wave of discoveries which propelled the race forward in the second leg of the relay, what poet Samuel Taylor Coleridge (1772–1834) described in 1819 as the second scientific revolution. This new wave of science and optimism included such figures as John Dalton (1766–1844) who was a pioneer in establishing chemical elements and in the laws of thermal expansion of gas; Humphrey Davy (1778–1829) who advanced accurate measures and weights and gave science a greater understanding of the fundamental processes of combustion; Michael Faraday (1791–1867) who made major advances in the understanding of electromagnetic forces, and was the inventor of the electric motor and transformer; William Herschel (1738–1822) who perfected telescopes for astronomical observation, discovering, in 1781, the planet Uranus, and also showing that our galaxy, the Milky Way, was only one of many; as well as Charles Lyell (1798–1875) the founder of the science of geology.

By then close attention was paid to how science should be properly done. In 1849 one noted scientist, John Herschel (1992–1871), son of William Herschel, wrote a book titled *Manual of Scientific Inquiry* in which he explained and promoted the new scientific method. His plan consisted of three steps: data gathering through observation and experimentation, the generation of hypotheses from the data, and then testing them to see if they were correct; a method that led to rapid advances in all branches of the physical sciences.

No doctrine or philosophy, or any other way of knowing, can become and remain a coherent body of growing and useful knowledge without a social context. In this regard biologist David Hull has observed that science is a process, characteristic not of individuals but of communities. Individuals, he says, rarely refute themselves, but fellow scientists are always eager to expose their colleagues' hypotheses to severe tests within a community of scientists in contact with one another and subscribing to a common set of principles and procedures (Hull 1988). It is within such social contexts that science is practiced and in which major changes, or as Thomas Kuhn puts it paradigm shifts take place. A paradigm shift is the replacement of a dominant theoretical model (the paradigm) for a more productive one. Such a shift becomes complete only when individuals who defend the old paradigm die off leaving the field to individuals who accept the new paradigm.

Science as ideally conceived and conducted is both a social process and a cultural system understood both epistemologically and historically. Moreover, it is so structured that in order to pursue one's research, the research of others must be taken into account. All this requires openness and constant communication among those who constitute scientific communities. Mechanisms for this process are professional scientific organization and professional meetings, journals and other publications. It is therefore no surprise to see such organizations and publications emerge with the First and their proliferation in the Second Scientific Revolutions.

Informal meetings in the seventeenth and later in the eighteenth centuries were held by various groups to discuss ideas of the day. One of those groups became the Royal Society of England. It started as a circle of specialists who began meeting in 1644 to discuss the new natural philosophy and to promote the knowledge of the physical world which that philosophy was generating. The name Royal Society first appeared in print in 1661, and in 1662 it was granted a royal charter to publish its proceedings under that name. The society's transactions consisted of weekly meetings to discuss scientific topics and to witness experiments.

At the end of the eighteenth and in the early part of the nineteenth centuries a number of other scientific institutions and "philosophical societies" sprang up whose purpose was the promotion and expansion of scientific knowledge. Among them were the Geological Society in 1807, the Astronomical Society in 1820, and the British Association for the Advancement of Science in 1831. The movement also expanded beyond professional organizations to the wider public. The most prominent was the Royal Institute founded in 1799 with the purpose of directly engaging the public in the form of lectures and experiments.

The early nineteenth century became the age of public lectures and public laboratory experiments which served the purpose of education, as well as entertainment, for an increasingly literate and aware population. The most famous of those lecturers, which still take places, is the Christmas Lecture established by Michael Faraday. That was also the time when the introductory textbook made its appearance and when science was first taught to children. It was also within that milieu that issues raised by science reached the level of general public debate. All this, together with the achievements of science and its promise, contributed to the creation of what Richard Holmes calls a new "secular philosophy of life," one in which "the wonders of creation (whether divine or not) were valued for their own sake" and shared by a wider more engaged public (Holmes 2008: xix).

Something new, however, was added to the scientific enterprise; that something was an "imaginative intensity and excitement" brought to science, as it was also brought to literature and the arts, by the Romantic

Movement (Holmes 2008: xvi). Romanticism was in part a back-to-nature movement, in part a look at earlier times, as well as new sympathetic attention to the traditions of humble people. It was also the pull of the exotic, and a taste for the bizarre, the untried, and the adventurous (Berlin 1999). Both the Second Scientific Revolution and Romanticism were parts of the spirit of the time, a dual spirit whose constituent components clashed in places but merged in others to give the rational nature of science its efficacy as well as a sense of creativity, excitement, and, at times, daring. It was partly that duality that enhanced the status of science in the popular mind and which in no small measure inspired many of those who chose to make it their profession. A central metaphor of that dual spirit was the exploratory expedition, whose mission was scientific and whose venues were the remote and still tantalizingly unknown regions of the world.

THE AGE OF SCIENTIFIC EXPLORATION

The wider world had opened up to Europeans at the end of the fifteenth and throughout the sixteenth centuries with the first passage around the Cape of Good Hope to Asia, the discovery of America, the first circumnavigation of the earth, and subsequent voyages of discovery. Reports were coming in from various sources as to the nature of the world outside Europe, and Europe was paying attention. Like their classical predecessors Europeans of the time saw a world beyond their own civilization, a world with other civilizations and barbarian tribes, peoples who spoke different, unknown languages. Europeans felt it necessary to classify and to name those different peoples, and to divide them into groups and to describe them; to bring order and system in understanding the human sphere as they were in understanding the world of plants, animals, seas, landforms, and climates around the world.

The first writer to bring to England an account of the world beyond the sea was Sir Walter Raleigh (1552 or 1554–1618), followed by the writings of Richard Hakluyt (1553–1616) whose books on discoveries, voyages, and with dealings in the Americas (1582, 1589–1600) did much to popularize the subject and to encourage settlement in the New World. Shakespeare and other authors at the time found his books a source for their own writing, and later writers on history and geography looked to his work as an example.

In the latter part of the seventeenth century, with the publication of his journals, the exploits of William Dampier (1651–1715) came to be widely known in England. He was the first Englishman to circumnavigate the world three times. On the basis of his first book *Voyages and Discoveries* (1699), the Royal Navy gave him command of an expedition to explore some of the western parts of Australia. Joseph Banks, one of the botanists on Captain Cook's expedition to Botany Bay, consulted Dampier's notes on the flora and the fauna of the area, and Charles Darwin and Alfred Russell Wallace took note of his work. He also contributed to the art of navigation by assembling and publishing in 1705 data on winds and currents and tides in oceans around the world, reports that were later used by Lord Nelson and James Cook.

It was also through his writings that such words as "sub-species," "avocado," "barbecue," and "chop-sticks," among many others, found their way into the English language. And his tales of tropical places and the strange things he had seen there sparked new interest in the wider world, and provided material for writers. For example, the story of one of Dampier's shipmates who shot an albatross became the basis for Coleridge's *Rime of the Ancient Mariner*. The story of another shipmate Alexander Selkirk who was marooned on a remote island was apparently the basis of Daniel Defoe's *Robinson Crusoe*, and Jonathan Swift explicitly compares the central character of his novel *Gulliver's Travels* to William Dampier. The appeal of those tales, says Glyn Williams, "was widespread. Scholars, publishers, seamen and merchants, and casual readers after a good yarn were all attracted." His books "became a standard model for voyages sailing to distant parts of the world" (Williams 2013).

The thirst for exploring, for discovering and describing unknown places, plants, animals, and people increased during the Enlightenment period. Only Dampier's second voyage was undertaken for naval exploration. The other two were as a privateer raiding Spanish ports and Spanish and Dutch commerce on the high sea. The first voyage sent out for the sole purpose of exploratory and scientific purposes began nearly six decades after the publication of Dampier's last book. That voyage, 1768–1779, was the first of three which, over an eleven-year period, not only advanced knowledge of distant places, but also tested new methods and new instruments and left its mark on the imagination of the time.

The first voyage was sponsored by the Royal Society with the mission of observing the transit of Venus across the sun, a mission accomplished on the island of Tahiti, followed by a search for a purported southern continent. James Cook (1728–1779), navigator, explorer, cartographer, and officer in the Royal Navy, was appointed to command that expedition. Cook and his crew were the first Europeans to contact the east coast of Australia, returning to England via Cape Horn, passing through the South Atlantic, mapping and surveying along the way and laying claim for Britain to South Georgia Island.

The second expedition, also under the command of Captain Cook, took them far to the south. In 1771 they crossed the Antarctic Circle, the first Europeans to do so. Due to a shortage of provisions they had to turn back before reaching the Antarctic continent. Cook's third voyage was undertaken in 1778 in search of the Northwest Passage. He proceeded to Tahiti, then north where he discovered the Hawaiian Islands, which he named the Sandwich Islands after the First Lord of the Admiralty. From there he explored the west coast of North America, finally returning to Hawaii where, in 1779, he was killed in an encounter with the natives.

Captain Cook published accurate maps of the places he had visited, and on the second expedition he helped the expedition's astronomer Charles Green to prepare nautical almanacs. It was also on that voyage that he proved the efficacy of the marine chronometer used for accurately establishing longitude. Another first of the second voyage was the successful prevention of scurvy among the crew by careful attention to nutrition. For this, Captain Cook was awarded the Copley Medal. Because of the success of his voyages he was made a Fellow of the Royal Society.

His expeditions also produced valuable collections of plant species and a collection of artifacts used by native people at the time of first contact. Captain Cook hypothesized that the people he encountered in the Pacific were of Asian origin, a hypothesis later confirmed with the study of the Malayo-Polynesian language family. The first expedition also produced drawings of Tahiti and vivid descriptions of its people and their customs, as did the other voyages. It was those graphic descriptions of the Pacific that turned a remote place into the romantic South Seas of Western imagination.

Captain Cook was popular in England for his exploits but the figure who most caught the public's attention was one of the botanists on the first expedition, Joseph Banks (1743–1820). Besides collecting and classifying plants, Banks also learned the language of the Tahitians and became better acquainted with them through interaction on a daily basis. He thus gradually became an ethnographer recording customs, facts about their clothing, their fishing methods, and cooking, their woodcarvings, religious beliefs, and, most fascinating to the public in England, their sexual practices. In this way, Banks introduced England to the wonders of a world beyond their shores, for which he became famous.

Banks's most important contribution, however, was his support of science and of scientists. He served as the president of the Royal Society from 1778 to 1819 which he managed with great efficiency. His contribution is perhaps most notable in the way he could spot talent and bring people together in the interest of science. In that regard he has been described as a "wonderful networker" (Kryza 2006:13).

Captain Cook's voyages of discovery were only the first. Others followed in the same spirit. The most extensive undertaken was the Challenger Expedition which the Royal Society sponsored ninety-four years after Cook's third voyage (1872–1876). The expedition was named after the scientifically rigged vessel which the Royal Society of London obtained from the Royal Navy for that project. Following in the scientific tradition of Captain Cook, the Challenger Expedition examined the nature of the ocean—the chemical composition of its water, its depth, temperature, circulation, specific gravity, light penetration as well as the physical characteristics of organic life at different depths and on the sea bed. The expedition's scientists also cataloged more than four thousand previously unknown species. The report of the findings of the expedition is contained in fifty volumes which laid the foundation for marine science. The space shuttle *Challenger* was named in their honor. The expedition's instructions also required that "Every opportunity should be taken of obtaining photographs of native races to scale." In compliance with those instructions hundreds of photographs were made and are now housed in the British Museum of Natural History (Dutton 1999: 75).

One still unknown place that beckoned was the "dark" continent to the south; the vast expanses of Africa whose fringes had been known to Europeans since the late fifteenth century, and whose coastal stations provisioned passing ships and provided traders with slaves. Yet the interior was still an enormous blank space on the map. Joseph Banks regarded this lack of knowledge as one of the failings of the age of Enlightenment. Europeans had sailed around the world on missions of discovery and exploration, yet they knew less about the interior of Africa just to the south than had the ancient geographers of Greece and Rome. Among the mysteries of Africa was the fabled "city of gold," Timbuktu, which had been reported in earlier writings but which had not yet been visited by Europeans.

In 1778 the Association for Promoting the Discovery of the Interior of Parts of Africa was founded, marking the effective date of African exploration. No European had ever seen the Niger River and no one knew the direction of its flow. The Africa Association had sent Major Daniel Houghton to locate the river and its course "as well as the various nations that inhabit its borders." But Houghton died in the Sahara without completing his task. He had, however, penetrated deeper into the interior than had any other European before him.

In 1795 Scottish physician Mungo Park (1771–1806), with the support of Joseph Banks, was selected to complete that mission. Park, who had sailed as ship's surgeon on an East India Company vessel to Sumatra, impressed the association with his knowledge of the Hadley quadrant which would allow him to make necessary observations, and enough ability as a geographer to find his way in unfamiliar places, as well as some acquaintance with the subject of natural history. In May of 1795 he set sail for Africa. Then from the mouth of the Gambia River he headed inland along Houghton's route. After close calls and enormous hardships he was finally able to reach the Niger. He was the first European to explore its central course and to record that the river flowed east. He also sought Timbuktu but was unable to reach it.

Park's account of his expedition delighted the Africa Association and thrilled the public. He became an instant hero, and the membership of the association dramatically increased. In 1803 he undertook a second voyage to Africa and once again to the Niger, this time with a more fully outfitted expedition including soldiers. The venture proved disastrous, with the white men dying of disease and Park himself killed in an encounter with tribesmen in what is now Nigeria. He may have seen Timbuktu on that trip but did not survive to report it. Park's expeditions however provided the greatest knowledge of Africa to that point.

The Africa Association was primarily interested in expanding knowledge for its own sake, but was secondarily interested in possibilities there for British commerce. Park, however, was solely interested in exploration. He was indeed "the first white man to the forbidding interior of Africa for the sole purpose of finding out what lay there, and to come back alive." He was lionized by English society and his book, *Travels into the Interior Districts of Africa* became a bestseller. In the process "he invented a new and glorious calling, creating an adventurous species of hero; the lone, brave African explorer" which captured the popular imagination, a figure "that fed the fantasies and filled the literature of Europe" (Kryza 2006: 20).

ALEXANDER VON HUMBOLDT AND THE INTERCONNECTEDNESS OF NATURE

Another key figure in the late eighteenth and early nineteenth centuries' expansion of science was Alexander von Humboldt (1769–1859) who Charles Darwin described as "the greatest scientific traveler who ever lived." Humboldt explored the heavens, but much more the earth and its varied forms of life. From 1799 to 1804 he traveled extensively in Latin America mapping the Orinoco River and its tributaries, eventually establishing its point of confluence with the head waters of the Amazon; observing the Pleiades meteor showers and the transit of Venus in southern latitudes; investigating the fertilizing capacity of guano (found in abundance off the Pacific coast of South America); and describing the great south-to-north flowing Pacific current which markedly affects the west coast of the continent, and which today bears his name.

He also contributed to climate science by inquiring into the origin of tropical storms and by suggesting the use of isothermic lines as a way of comparing climatic conditions on different continents. In fact he is arguably the founder of the science of climatology. He also discovered the decrease of intensity in the earth's electromagnetic field, and was an earlier contributor to, and an important influence on, the study of the geomagnetic field. Also his account of the propagation of seismic waves became the basis of modern seismology.

Humboldt's principal interest, however, was the biodiversity of the world (he sent back some 60,000 botanical specimens from his expeditions) and he described the relationship of that diversity to the physical environment. His methods were those of science (systematic observation and quantification) and his vision was that of the interconnectedness of nature which, despite the apparent chaos of diversity of life forms, offered the possibility for the kind of systematization that Newton had worked out for the universe. Humboldt's holistic view of nature meant that he also considered culture in relationship to the environment. For example, he asked why the tropical zones rarely gave rise to civilization.

In 1804 Humboldt published an account of his expeditions in a book titled *Personal Narrative of a Journey to the Equinoctial Regions of the New Continent*. The book was widely read and influenced a new generation of scientists. As noted above, rational science and Romanticism merged in an important way during Humboldt's time, but they also clashed. One point of contrast was the position held by some Romantics that we should understand nature intuitively without resorting to the scientific method. Experimentation, they said, was interference with nature, and the formulation of natural laws was destructive to man's feeling for the wonders of nature. Humboldt argued quite the contrary; that an understanding of nature through the methods of science would not only enhance our knowledge, but that it also represents the development of a natural human urge, which he said he had observed even among the most "primitive savage nations," a feeling which he described as the "terror-stricken sense of the all-powerful unity of natural forces" which gave rise to religion and philosophy and from there to knowing through the empirical sciences.

There is also an aesthetic aspect in that relationship seen in "the universality of Man's enjoyment of nature" for "communion with nature, mere contact with the free air, exercises a soothing yet strengthening influence on the wearied spirit, calms the passion, and softens the heart when shaken by sorrow to its innermost depth." Everywhere, he says, "in every region of the globe, in every stage of intellectual culture, the same sources of enjoyment are alike vouchsafed to Man" (Humboldt 1849: 3).

Among the goals of his writings were the presentation of "the physical description of the universe, considered as a distinct science" as well as the examination of the "influence of the external world on the imagination and feelings which has acted in modern times as a powerful impulse towards the study of natural science" (1849: 37–38). Those tasks, he says, and the grandeur of nature herself, call for an elegant style of presentation since "the profound conceptions and enjoyments she awakens within us can only be vividly delineated by thought clothed in exalted forms of speech" which are "worthy of bearing witness to the majesty of creation." Thus "a certain degree of scientific completeness in treatment of individual facts is not wholly incompatible with a picturesque animation of style." In his writings and his explanation of why he wrote the way he did, we see a prime example of that dual character of his time; empirical, rational science infused with the adventure of travel and discovery and knowing and of eloquence in presentation.

This was the age of the public lecture and Humboldt was one of its star attractions. He presented a series of lectures in Paris on the nature of the physical world, and as a member of the Berlin Academy he was permitted to lecture at the University of Berlin. Between November 1827 and April 1828 he delivered a series of sixty-one lectures there, to halls packed with students and professors of all faculties. Interest was so great that he repeated the series to the wider public. This time he spoke to capacity crowds in a concert hall, to people from all walks of life; the royalty in the boxes and ordinary citizens below and in the orchestra pit. Women were highly represented among the audience of thousands which attended the lectures. "Never had there been such a socially mixed audience in the city's history" evidence of thirst for the kind of knowledge which science could provide (Botting 1973: 234).

Beginning in 1845, Humboldt published the contents of his lectures in several volumes over the next twenty-five years titled *Cosmos: A Sketch of the Physical Description of the Universe*. In those volumes Humboldt presented his view of the unity of nature and the scientific way of studying it. The first volume sold out in a matter of months. The final volume appeared posthumously in 1862, completed from notes he had left of the unfinished work. His knowledge was voluminous, but he sent manuscript copies to experts in various fields for confirmation and further comment. He also brought to his technical descriptions a literary style which sustains the reader's interest, what has been described as a unique blend of poetry and science (Wulf 2015), liberally illustrating with examples from around the world, the Himalayan Mountains, the Mississippi Valley, the South Seas, and in his explanation of earthquakes, volcanoes, climatic zones, and so forth. His work was widely read by academics and laymen, and on the basis of his writings he became a world figure. For example, Thomas Jefferson invited him to Monticello for several weeks as a "consultant."

A SENSE OF UNITY OF NATURE AND THE SCIENCE THAT STUDIES IT

The interrelationship of life forms that Humboldt emphasized was already a topic of speculation and investigation. The word "ecology," which labels what would become its own branch of science, was coined by Ernst Haekel (1834–1919), a physician, naturalist, and biologist who described and named thousands of new species and who designed a genealogical tree including all life forms. Among the various words that Haekel coined, which today form a part of the scientific vocabulary, are "stem cell," "phylum," and "phylogeny." His work set on a firm foundation the current view of nature in its interconnectivity and the productive scientific way of describing and explaining it, a view which was within the inclusive vision that Humboldt had made popular.

Another example of the recognition of the unity of reality and the interconnectedness of the sciences is seen in the work of William Whewell (1794–1866), an influential figure in nineteenth century Britain who wrote on many subjects, such as geology, astronomy, mineralogy, mechanics, political economy, theology, moral philosophy, international law, architecture, education reform, as well as the history and philosophy of science; varied interests from which came his thoughts on the interconnectedness of science (Wilson 1998). He was also in close association with men of science in his day, and his influence was acknowledged by John Hershel, Charles Darwin, Charles Lyell, and Michael Faraday, men who frequently turned to him for philosophical and scientific advice. Whewell also wrote of holism in science, what he called in his *Philosophy of the Inductive Sciences*, "consilience," literally a "jumping together" of knowledge by linking facts, and fact-based theories gained from across disciplines to create a common ground of explanation.

However Humboldt's vision of the interconnectedness of nature and the web of life in which humans are a part, as well as similar views such as those of Whewell, lost their appeal as science became more compartmentalized, and in the English speaking countries as the idea of science itself narrowed to its present meaning (see chapter 5).

THE ORIGINS OF PHYSICAL ANTHROPOLOGY

German physician, physiologist, and naturalist Johan Blumenbach (1752–1840) was one of those scientists in the latter half of the eighteenth and the early years of the nineteenth centuries whose interests ranged across the sciences and who was interested in ideas of evolution, creation, the nature of life, and life forces. One of the topics which attracted his attention was the question of the physical nature of life, how it differed from inorganic matter in terms of procreation, nourishment, and reproduction on the physical level, on another level its characteristics of self-development and self-defense. Blumenbach's theory to account for those differences between the organic and inorganic centered on what he called *vis formativus* or *Bildungstrieb*, "formative drive" which according to Blumenbach was not a chemical or mechanical quality but rather a force in its own right. He compared the uncertainties involved in his concept with uncertainties about gravitational attraction. His concept of formative drive was influential at the end of the eighteenth century as seen in the writings of Emmanuel Kant and others.

Blumenbach was also one of the first to identify the different species of primates, and to view humans in terms of natural history. He was also interested in human physical variety, and devised a racial classification which was used by scientists studying evolution in the nineteenth century. He is thus widely considered the first physical anthropologists. Based on skin color and anatomical features such as the shape of the cranium, he divided the species into five racial categories. They are *Caucasian*, white; *Mongoloid*, yellow; *Malay*, brown; *Ethiopian* (which included the peoples of Sub-Saharan Africa), black; and *American*, red. He believed in monogenesis, that is, that all races came from a single variety; according to his theory, from the Caucasian race. The differences among them, he said, came about due to differences in geography, diet, and mannerisms. He did not, however, believe in the inferiority of some races as did others at the time, and argued against those who did.

DISCOVERING THEIR OWN FOLK ROOTS

While traveling men of science were exploring distant places and contacting remote peoples, attention in Germany was turned toward classical and Oriental languages, as well as toward their own language, and toward a careful inward examination of their own folk roots, all of which converged on the study of Europe's nearer and distant past. This was a time when in Germany the Romantic Movement brought "a new intellectual world" into being, a world "over and against the world of the natural sciences, a world where nature is only the environment, and human activity the central fact" (Hodges 1969: 6). This clashed with the view that gave priority to the scientific method that had been worked out so successfully for the physical world. Out of this confrontation arose that duality which was derived from the Romantic perspective on the one hand and on the other hand from the rationalism of the Enlightenment which, as we saw above, characterized the spirit of the time. In Germany, the period between 1770 and 1830 was the time when both the rationalist scientific and the romantic movements merged and contrasted with one another, as they did across Europe, but in Germany in a different way and with greater intensity than elsewhere, in what one historian describes as "a period of concentrated intellectual activity such as has probably no parallel anywhere in modern history" (Hodges 1969: 5).

The impetus for this new way of looking at the past was Johann Gottfried von Herder (1744–1803), whose influence in nineteenth-century Germany was widespread, affecting the Romantic Movement as well as the growth of nationalism, while introducing new concepts with implications for anthropology. Herder believed that nations differed from one another in very significant ways. Each nation, he said, has its own unique spirit, its *Volksgeist*, that had developed in its own environment and according to its own history. Each national spirit mentally constructed social life and animated the nation of which it was a part. That unique spirit was expressed and passed on in the form of the way of life and the language of the people, transmitted down the generations by word of mouth and by customary example.

Herder also believed that urban society, with its Greco-Roman and Christian roots, the universalizing civilization of his time, had suppressed this more genuine level of the nation, a level which still survives among the humble peasant class and among the less sophisticated people of the towns and cities. Thus another way of looking into the past was opened up through the systematic study of the contemporary manners, the speech, the customs, the tales, the songs, and crafts and arts of the people just outside of town and also from the humbler classes of urban society. Herder advised his countrymen to collect folk poetry, and he himself edited a book of German folksongs.

Scholars took him up on it. Among them were Wilhelm and Jacob Grimm who in 1812 published the now famous *Kinder-und Hausmärchen* "Children's and Household Tales," which presented in literary form the fairy tales we all know so well. Folklorist Richard Dorson says that "few books in the history of printing have enjoyed such commercial and intellectual success." Dorson calls their work an act of intellectual discovery which opened up, for the educated classes, a window "into the submerged and despised culture of the peasantry, disclosing undreamt of riches in such an outwardly barren environment" (1966: v). Between 1816 and 1818 the Grimm Brothers published *Deutsche Sagen*, "German Sagas" a collection of legends to complement the fairy tales of their first book, and in 1835 Jacob Grimm published *Deutsche Mythologie* "German Mythology."

Their motive was to reach back beyond the accretions of Classical and Christian civilization, in an effort to reconstruct the original Teutonic culture which they said forms the core of the true German nation, its formative *Volksgeist*. To them, folktales and legends were flotsam of an early culture surviving in simple folk societies. That flotsam could be assembled and used to reconstruct ancient Germanic mythology, somewhat like an archaeologist collecting and reassembling shards of some ancient urn, filling in the gaps with inference where needed.

Their method was to take the tales and legends directly from the lips of the storytellers, a method which Dorson says led "deep into a vast subterranean narrative lore of the unlettered peasant" (1966: v). Dorson calls this a new approach, a science that found an immediate place within the intellectual pursuits of the time. The label the Grimm Brothers chose for this new line of investigation was *Volkskunde*. In 1846 William Thoms in England, following the German example coined the term "folklore" from two English words which clearly expresses the nature of the discipline, a term which has diffused into other languages as the discipline spread.

The Grimm Brothers's claim that they had collected the stories from the folk stratum of society has since been proven false. Their real sources were from the literate middle class (Ellis 1983). Yet the model they established became that of subsequent work in folk communities, and that kind of close face-to-face elicitation of material culture as well as beliefs, customs, and oral traditions is characteristic of ethnographic work among exotic peoples, and is a standard part of ethnography today.

EXOTIC LANGUAGES AND THE SEARCH FOR A DISTANT PAST

Language was another way of reaching into the past. A hallmark of Renaissance scholarship had been the attempt to reconstruct the Greco-Roman past by discovering, collecting, collating, and analyzing ancient texts, as well as by examining the material remains left by the ancients. Their goal was to reconstruct what they thought was the highest pinnacle reached by their civilization. This effort also had a contemporary purpose, namely to improve their own societies and culture by imitating what they considered a higher stage of civilization, a highly sophisticated way of life which had come before them and which had since passed away, but whose traces were found in the ancient art and architecture around them and especially in ancient texts.

Another linguistic avenue of studying the past—not just the past of a single nation or of classical civilization but of wide the sweep of history—was opened up with the discovery of Sanskrit, the classical language of India, a language analogous to Greek and Latin in the West, with its body of literature and its rich philological tradition. Leonard Bloomfield says of that discovery, which included precise grammatical writings, that it provided Europeans for the first time with a complete and accurate description of a language based on observation and not just theory.

Moreover, it provided Western scholars with a linguistic description similar in accuracy to those of Greek and Latin, descriptions which exhibit similarities between the three classical languages, a fact which, in those inquisitive times, invited comparison. That is what Sir William Jones (1746–1794) did, when in 1786 he showed the strong grammatical similarities between the three languages, as well as similarities among them and Persian and Gothic (an earlier Germanic language). Jones argued that those distinct but similarly structured languages must have had a common source that lay far back in time. This discovery set off a new wave of research, first to reconstruct through systematic comparison the proto-language of what would become known as the Indo-European family of languages. Bloomfield says of that enterprise that "in spite of poor equipment, and thanks to the energy of their workers, the historical and comparative study of Indo-European languages became one of the principal enterprises, and one of the most successful, of European science in the nineteenth century" (1933: 12).

Sanskrit studies were further advanced through the work of Friedrich Schlegel (1772–1829) with his book *Über die Sprache und Weisheit der Indier* (1808) ("On the language and Wisdom of the Indians"), along with other writings, and through the work of his brother August Schlegel who established Sanskrit studies in Germany. The first systematic comparison of the Indo-European languages was presented in a treatise by Franz Bopp (1791–1867) which examined the inflection of verbs in Latin, Greek, Sanskrit, Persian, and Germanic. In 1818 Danish scholar Rasmus Rask (1782–1832) showed that the sounds of words in the Germanic languages bear a regular and formal resemblance to those of other Indo-European languages; a fact that was more fully elaborated in 1822 by one of the Grimm Brothers, Jacob Grimm, in volume two of his comparative grammar of the Germanic languages titled *Deutsche Grammatik*. Grimm presented a systematic exposition of the correspondence of consonants between the Germanic and other Indo-European languages which became known as Grimm's Law, a work which Bloomfield says remains unrivaled today.

That regularity suggested that linguistic change in general, which had already been detected in written records, might obey the same kind of laws that physicists had unveiled for the material world. There were, however, still some exceptions to Grimm's findings. Hermann Grassmann, in what came to be known as Grassmann's Law, accounted for some of the differences by showing that they resulted from developments in each of the languages. Later Karl Verner solved the problem by explaining seeming exceptions to Grimm's Law by showing the regular way the sounds change within the phonetic environment of the word. In other words Grimm's Law proved that sound laws exist and Verner's Law showed that they work.

In the latter half of the century the focus in linguistics was on history, and the model was that of biology, with language viewed in organic terms, with daughter languages descending from mother languages, whose relationship and paths of divergence could be systematically described in the form of family trees. Traces of earlier stages of the language were still present in the language of the times, and its representatives could be found for earlier times in the written record. Thus like Georges Cuvier, who reconstructed extinct animals from fossil remains, linguists set about reconstructing those earlier pre-literate stages of language, and through the comparative method to reveal the genetic relationship of the languages they studied.

The recognition that there was regularity in language change led to confidence in the discovery of *Ausnahmslosigkeitsgesetze*, laws that operate without exception, a view that proved overly optimistic, soon leading linguists to account for language variability and change by examining not only patterns of internal change through the written record and the comparative method, but also by reference to its geographic and social distribution. This turn in direction initiated another wave of field work with investigators traveling from place to place across the countryside recording the speech of people on the folk level of society and later examining language use in social context, resulting in the field of dialectology and eventually sociolinguistics.

Andre Gingrich, viewing the late Enlightenment's and the Romantic period's interest in oriental languages, believes that those innovative philological methods in Germany and their results had "lasting effects in the form of solid empirical description of ethnographic and cultural variation, as well as intense mastering of foreign languages" (and we might add attention to the inner structure and the classification of languages) which "set a standard for a future international anthropology" (2005: 70).

LANGUAGE, THOUGHT, AND CULTURE

Another development at that time was the introduction of the idea of the organic relationship between language and national culture. Here again we see the source in the writings of Herder. In Herder's view each group forms a distinct entity with its own history and its own animating spirit. Language plays a key role in this process not only as an instrument of communication but it also acts as a form of thinking and a treasure house. Language is a treasure house because knowledge and experience are accumulated in the expressive resources of the common language, and language is a way of thinking since that knowledge is transmitted to children as they grew up in the community, thus the common store of knowledge is passed down from one generation to another via language.

It is also language that fixes experiences and falsehoods in the collective consciousness. In that way language and its cultural content help individuals and the community make sense out of the world in a consensual form, and in the process molds the vision of the world in which they live. In this way language determines the boundaries of human cognition and acts to shape the content within those boundaries (Schaff 1973: 1–25). In the first half of the nineteenth century, after the discovery of Sanskrit and the work of Franz Bopp, Jacob Grimm, and others, we see the development of what would become modern linguistics.

A prominent scholar in that tradition was Wilhelm von Humboldt (1767–1835), the brother of naturalist Alexander von Humboldt whose theoretical position and empirical research would become a part of linguistic anthropology. Humboldt paid close attention to linguistic variation saying that the model of Latin grammarians should be modified when studying different languages, to confirm to their individual nature. Humboldt also saw language not just as a static structure of grammar and lexicon as it is perceived in philological study. Instead he articulated both a structural and function view of language in which each language consists of *ergon*, a mechanism, or as he put it elsewhere, as the skeletal organization of language, and a dynamic creative force which he called *energia*, a constantly functioning action of the mind which mediates between external reality and internal thought.

Language, he said does not simply reflect reality, but also helps shape the comprehension of reality for individuals and for communities and to make possible expression in speech. This coordinating force he called the inner form, an activity of the mind in linguistic comprehension and communication. Following Herder, Humboldt saw an intimate relationship between language and the nation, each different in its own individual manner and each of those distinctive ways characterized by the way the language influences the world view of the community.

REIMAGINING TIME

The first decades of the nineteenth century saw the emergence of geology, a new science which not only greatly increased knowledge of the earth's composition and form, but also revolutionized scientific thought by greatly extending the time span. According to the traditional world view, time began with Creation as described in the Book of Genesis. In 1654 the scholar and bishop of Ireland, James Usher, examined the genealogy of the Old Testament and placed the date on the twenty second of October 4004 BC, a reasonable assessment in an age when the Bible was considered the source of unquestioned truth, at least when it came to religious matters, and for many when it came in conflict with acceptable science. Thus the commitment to the biblical chronology continued into the eighteenth century and in some cases into the nineteenth century.

One notable example was George Rawlings, a highly regarded and prolific historian of the ancient history. Rawlings was one of the proponents of changes that had taken place in "the whole world of profane history," changes which reflected the times with insistence on the careful evaluation of sources, allowing for the consideration of reported events on separate levels of "historical value" and thus the writing of more accurate accounts of the past. To that end Rawlings advocated certain rules of evidence in order to distinguish between historical events and mythic narrative. The result, he wrote, was that "the views of the ancient world formerly entertained had been in tens thousands points either modified or reversed." When it came to the literal truth of the Bible, however, he held firm to the literal interpretation.

In his 1859 Bampton Lecture (the year that Darwin's *Origin of Species* was published), Rawlings titled his talk "The Historical Evidence for the Truth." He said that "We possess in the Pentateuch [the first five books of the Old Testament] not only the most authentic account of ancient times that has come down to us, but a history absolutely and in every respect true" (Greenslade 1963: 259–260). Thus Moses's mother Jochabed had probably met Noah's son Shem, and that Shem was probably acquainted with Methuselah (Daniel 1962: 61).

Long before this, however, a problem with the biblical chronology had appeared, for as more and more was learned about the nature of the earth, the shallow biblical time depth could not accommodate the discovery and evaluation of the phenomena which had been observed. One problem was how to account for the fossils found in different strata of the earth's surface. For centuries people had uncovered fossils in rocks, but it was only in the late eighteenth century that their nature came to be understood when French naturalist and zoologist George Cuvier (1769–1832) convincingly argued that fossils were actually petrified animals which are now extinct. He also noted that the fossilized remains of different kinds of animals were found in different strata of the earth, and concluded that those differences mark periods of catastrophic extinction followed by the rise of new species. Today Cuvier is known as the founder of biostratigraphy. Although Cuvier's theory, known as catastrophism, involved a much longer time span than that calculated by Bishop Usher, it did not logically clash with the biblical interpretation, for the Bible also notes one major catastrophe, namely the Great Flood. In fact, it was common in those days to refer to the early geological past as the antediluvian period.

The developing science of geology was deepened with the work in Germany of Abraham Werner (1749–1817) who had been examining different kinds of rocks retrieved from mines. He noticed that the hardest rocks contained no fossils, while other rocks did. He took that as a clue as to their age, and on that basis he devised a system of classification. In England, William Smith (1769–1839) was employed in the construction of a system of transportation canals during the pre-railroad era. During construction he observed different strata beneath the surface, and in 1815 he published the first geological map of Great Britain which covered England, Wales, and part of Scotland, earning for him the unofficial title of the father of English geology. Thus a greater understanding of the strata of the earth, and the variation over time of the kinds of fossils which were found embedded within them, opened up the possibility of reading the earth's history and with it the suggestion of much greater time spans than given in a literal reading of the Bible.

Questions also arose about the surface of the earth, its river valleys, mountains, coastlines, and so forth, and why, for example, sea shells were sometimes found on mountain tops far away from the sea. One possibility was that land forms were the products of natural forces, the kind that were still active and observable, such as volcanoes, floods, and erosion. If the time span were long enough than the one then assumed, those processes would explain both the fossil record and the geological data. Charles Lyell (1798–1875) brought it all together within his highly influential *Principles of Geology* (1830–1833) which offered a much longer time span than had been previously assumed, thus opening the way for further advances not only in geology but in science in general, most notably with the theory of biological evolution.

BIOLOGICAL EVOLUTION

Between 1831 and 1836, Charles Darwin sailed on another round the world voyage of exploration as guest naturalist on a British naval vessel HMS *Beagle*. During the voyage he made systematic observations of species of plants and animals in their natural environments. In 1859 after long contemplation and further study, he finally published the results of his work titled *The Origin of Species* in which he made the case for biological evolution. There was resistance from religious quarters, but his theory was widely accepted in the scientific community, for like Newton, Boyle, and others, Darwin's theory was based on a solid body of empirical evidence, and it conformed to the pre-existent belief in immutable laws, in this case regularities in the process of evolution, which are much like the laws that underlie Newton's celestial mechanics. Experiment was also part of the scientific process, and though Darwin's work was observational in nature, he also tested his ideas through experimentation with the breeding of pigeons. His theory was also compelling and appealing in no small measure thanks to the sheer adventure of the famous voyage of the HMS *Beagle* which followed the wake of Dampier, Captain Cook, and Alexander von Humboldt.

Humboldt's vision of the interconnectedness of the natural world also played a role in the conceptual framework which made it possible for Darwin to interpret the vast array of plant and animal life in the way he did. In a letter to a friend in 1831 shortly before he reached Brazil on HMS *Beagle*, Darwin wrote that Humboldt was "like another sun who illumes everything I behold," and later in his autobiography he wrote that Humboldt's account of his journeys "stirred up in me a burning zeal to add even the most humble contribution, to the noble structure of Natural Science." Nor would his theory have proven so credible if he had not adhered to the methods of inductive science which had proven so successful in other fields as explained by John Herschel whose book he had read.

Not only was the conceptual framework for Darwin's theory shaped by the science of his day, but the idea of evolution itself was already in the air. The first to propose a full theory of evolution (1801) was French naturalist Jean Baptiste Lamarck (1744–1824). He started as a botanist but was also engaged in vertebrate and invertebrate zoology. He also knew the fossil record that was increasing at the time. And he was active in the scientific classification of species. The evidence for his theory came from those fields and such features as vestigial structures of organisms as well as selective breeding. In 1801 Lamarck proposed a full blown theory of evolution in which he maintained that traits acquired in one generation could be passed down to succeeding generations in the process of adapting to the environment in which the species lived. The greater the use of an organism said Lamarck, the longer the life time of the individuals, and as the organism adapted to its surroundings it became ever more complex over time. For example, giraffes evolved ever longer necks in order to feed on higher leaves on trees. Lamarck's conjectures were grounded in the scientific method and carried out on the foundation of his encyclopedic knowledge of botany, vertebrate, and invertebrate organisms and the fossil record.

Lamarck was followed by French naturalist Geoffroy St. Hilaire who proposed a different version of evolution in the 1820s. And in 1844 Robert Chambers of Edinburg published a version of the theory of species mutability which was a best seller. In the far-off Dutch East Indies, naturalist Alfred Russell Wallace (1823–1913), during a bout of fever, also saw how evolution could account for the evidence available. It was Charles Darwin, however, who formulated a theory which could account for this elaboration, along with a mechanism—natural selection—to explain how it works. Darwin's theory thus came at a time when such a radical idea was acceptable, but by no means wholly accepted by all of science or by most of the public. But the theory could not have appeared as it did, in fact Darwin might not have thought of it at all, had it not been for the greatly extended time depth which Charles Lyell had introduced in his widely read work *Principles of Geology*. In fact Lyell's was one of the formative books in Charles Darwin's preparation, a copy of which he took with him on his voyage.

CONCLUSION

In sum, the seventeenth century was characterized by a great movement forward in mathematics and the physical sciences known as the Scientific Revolution. The eighteenth century was a time of optimism, a belief in progress and the assumption that the search for laws governing how things work could be expanded from the physical world to other areas. At the end of that century and during the first decades of the nineteenth century, Europe underwent a second scientific revolution which involved the fusion of rationalism and Romanticism to create a single movement with two branches, a movement nowhere more active than in Germany.

In the study of human activity, one branch of that movement studied different aspects of culture such as language, mythology, law, religion, poetry, and so forth. The methods employed were the comparative method and genetic analogy of biology. Its purpose was to show how each of those cultural systems were governed by an inner law which determined their general structure and the course of their history. The founders of this branch were Bopp, the Humboldts, W. Grimm, Savigny, among others. The second branch was associated with the Romantic Movement and was characterized by a sense of the exotic past and other cultures, an element also present in the study of language, myth, and religion. The focus of this second branch was the study of the life of nations, tracing their interactions as organized forces, with attempts to discover laws which determined their rise and fall and attempting to get inside them (Hodges 1952: 6).

It was in that intellectual upsurge, with its expanding range of scientific inquiry infused with strains of Romanticism, that anthropology began to emerge as an independent scientific enterprise. Anthropology today retains that dual legacy, accepting the unity of the species while seriously attempting to enter and understand the thought worlds of different cultures. In this sense, says historian of anthropology George Stocking, the discipline follows the German Romantic tradition which has had a major impact in shaping American anthropology (Stocking 1982: 408). The discipline we know today, therefore, represents that duality; the Enlightenment interest in universal features and the scientific method, while at the same time, remains fascinated by human variation and still responding to the lure of other places and other times.

TWO

European Origins in an Age of Science

HUMAN FOSSILS AND THE BEGINNING OF PALEOANTHROPOLOGY

Carl von Linné (also known as Linneus 1717–1778) devised a scientific system of classifying plants and animals based on a comparison of their physical similarities and organized according to genus and species. In 1758, Linneus took Man down from his unique status as a creature created in the image and likeness of God, and placed him alongside other mammals in the animal kingdom, a position which challenged the then prevailing view of Man in nature. For example, George Cuvier, who led the way in paleontology, adhered to the biblical view asserting that there were no human fossils since humans were created by God in the form in which he is today. This view began to change in the first half of the nineteenth century as speculations on evolution became more widely known.

It was in that context when in 1856 a peculiar skeleton was discovered in a quarry in the Neander Valley (Neanderthal) of West Germany. The bones attracted attention for their size and for other skeletal features which differed from those of modern man. The question was what kind of being did they represent? A local schoolteacher by the name of Johann Carl Fuhlrott who was fully involved in the scientific mindset of his time, was not slow to see their significance. He brought them to the attention of Hermann Schaaffhausen, professor of anatomy at the university in Bonn who three years earlier, in 1853, had written a paper in which he discussed the possibility of human fossils. Upon examining the bones which Fuhlrott had brought him, he decided that they were indeed fossils from an earlier period. In 1859, the same year as the publication of Darwin's *Origin of Species*, Schaaffhausen together with Fuhlrott published a paper in which they identified the bones as the fossil remains of a kind of people that had inhabited Europe at a very early time.

A skull discovered in Gibraltar in 1848, and later finds in Belgium, all had similar characteristics showing that the fossils found in the Neander valley were not unique, and that they represented not a deformed modern man as had been argued by some skeptics, but that they were indeed the remains of a different kind of man. In 1886, one of the Belgium sites yielded two almost complete skeletons which confirmed Schaaffhausen's conclusions. Also the antiquity of the finds was confirmed by their association with bones of ancient horses, deer, hyenas, and mammoths, as well as with stone tools. By today's reckoning the Gibraltar skull is some 50,000 years old, and those from Belgium date back 60,000 years.

Neanderthal Man attracted wider public attention when Charles Lyell wrote an essay in 1863 titled "The Geological Evidence of the Antiquity of Man." And in the same year Thomas Henry Huxley, a supporter of Darwin, wrote that the finds in Germany and Belgium were indeed proof of the theory of evolution. He said, however, that the fossils were not the remains of a "missing link" between apes and Man but that they represented an earlier stage in human evolution. Once the antiquity of Neanderthal Man was recognized, attempts were made to place him in the genealogy of the human species. William King argued that the bones appeared too primitive to belong to the species *Homo sapiens* and assigned them to the new category of *Homo neanderthalensis* a classification that has been put in doubt by later studies. But with the discovery of Neanderthal Man we see the starting point for the new science of paleoanthropology.

THE IDEA OF PREHISTORY

The first half of the nineteenth century was also the formative period for the idea of prehistory; the first time, says Glyn Daniel, that scholars saw prehistory as something that could be determined by means of archaeology. Prehistoric tools had often been found. But until the scientific period they were thought to have been fairy arrows, elfshot, or the product of thunderstorms. When the artifacts were recognized as manmade, the reasoning was that there must have been a time in Europe when there was no metal for that purpose, and thus stone tools, for if there had been metal then the tool makers would have used it for their implements. Also native peoples of the Americas and the Pacific used only stone tools since they had no metallurgy, therefore there must have been a similar time in European prehistory.

Christian Jorgensen used this concept in an effort to bring some sort of order to the growing number of artifacts in the Danish National Museum. He thus arranged the collection into three stages: stone, bronze, and iron.

He published his classification in the museum guidebook in 1836. And in 1849, J. J. A. Worsae made that three staged archaeological-based chronology more widely known through his book *Primeval Antiquities of Denmark*. This system of classification was later expanded by Robert Lubbock—an early figure in archaeology, ethnography, and biology—who went further and identified two phases of the Stone Age, the Paleolithic and the Neolithic (the Old and the New Stone Ages), which were distinguished by the greater sophistication of tools in the later period. Charles Lyell included in that chronology in the tenth edition of his *Principles of Geology* (Daniel 1962: 56–59).

In 1836 an amateur antiquarian Jacques Boucher de Crevercoeur de Perthes (1783–1868) "inaugurated the epoch-making advancement of prehistory" by arguing that the stone hatchets that had been discovered were contemporaneous with now extinct animals. This indicated, he said, that culture dated back not a few millennia but to a period prior to the geological present (Lowie 1937: 7). In 1846, he advanced his argument in *Treatise De l'Industrie primitive*. This created a problem, for as James Cowles Pritchard wrote in 1843 in his *Researches into the Physical History of Man*, "Many writers who have been by no means inclined to raise objections against the authority of the sacred scriptures . . . have felt themselves embarrassed by the shortness of the interval between the Noachic Flood and the Period at which the record of various nations commence, or the earliest date to which their historical memorials lead us back" (Daniel 1962: 62). With the increased acceptance of the extended span of geological time, that problem was solved, and scientists were more inclined to accept Boucher de Crevercoeur de Perthes's argument. One reason why some still rejected his theory was scientific skepticism. But in 1854, a former opponent examined the sites from which the evidence had been taken and announced that he was convinced of the antiquity of the artifacts. In 1858, British paleontologist Hugh Falconer agreed. Others followed including Charles Lyell who in 1859 announced his complete acceptance of the theory of the antiquity of prehistoric stone tools.

By 1865 the idea of the long prehistory of man was well advanced, and John Lubbock could write in *Prehistoric Times* that "The first appearance of man in Europe dates from a period so remote that neither history nor tradition can throw any light on his mode of life" (Daniel 1962: 61). He also wrote in the first edition of the book that "Of late years a new branch of knowledge has arisen, a new science has . . . been born among us, which deals with times and events far more ancient than any which have yet fallen within the province of archaeologists," a science which "forms a link between geology and history" (Daniel 1962: 63).

DIGGING UP THE HISTORICAL PAST

As geology had pushed prehistory back to the very remote past, archaeological investigations were examining and extending the historical past back to some five thousand years. The study of ancient history at that time was connected with the Bible and with the Greek and Roman classics, both areas of great interest in the eighteenth and nineteenth centuries. In the inquisitive scientifically oriented spirit of the time, attention turned away from just biblical and classical texts, to a search for physical remains from Egypt, Mesopotamia, and the Eastern Mediterranean, which might confirm what the written record revealed, and in the process expanding knowledge of those precursors of Western Civilization. Thus from around 1800 on, work on the material and intellectual life of ancient people extended the time horizon from two millennia to at least five thousand years, opening up for us "insights into the development of human life far wider in time and space" than had been the case until then (Friedrich 1957: ix).

Already in the eighteenth century, the study of history was beginning to include closer attention to historical archives and manuscripts of different kinds formerly overlooked by historians, as well as closer attention to ancient artifacts and archaeological and historical sites. Excavation began in the ruins of Herculaneum in 1738 and in Pompeii in 1749, Roman cities that had been covered by lava flows from an eruption of Mt. Vesuvius in 79 AD. Another step in the direction of more advanced historical archaeology, took place between 1798 and 1801 during a French military campaign in Egypt.

Napoleon Bonaparte dispatched on the military expedition, along with the troops, a five hundred man team of civilian scientists, to study the ancient Egyptian culture. Among them were specialists in biology, chemistry, and languages, the first ever overseas archeological expedition. With that team went Jean-Francois Champollion who deciphered the expedition's most famous discovery, the Rosetta Stone. The text engraved on the stone was written in the ancient Egyptian language in hieroglyphics and in the simpler demotic script, along with a Greek translation which provided the key to the Egyptian text. With this information, hieroglyphs no longer masked the written record and a window was opened onto the civilization of ancient Egypt.

Through the work of English and German scholars, another ancient linguistic monument was deciphered, the lengthy inscriptions in Behistun in Iran which were carved on the face of a cliff between 522 and 486 BC. The text was written in three languages, Old Persian, Elamite, and Babylonian (the forerunner of Akkadian), thus revealing the nature of those languages and making possible the reading of other texts. Paul Emile Botta (1802–1870) was a scientist who served for a time as French consul in Mosul in what is now Iraq. While in the country he initiated large-scale archaeological excavations. In the 1840s, believing he was digging in the ancient city of Nineveh, Botta discovered the palace of the Assyrian ruler Sargon II.

August Henry Layard was a diplomat, art historian, an expert in the cuneiform writing system and the language it recorded, as well as an archaeologist whose interest was in locating ancient biblical cities. He excavated several sites in present day Iraq, where in the 1840s he uncovered the cities of Nimrod and Nineveh and the Assyrian library of Ashurbanipal. The latter contained the text of the Gilgamesh Epic, the Akkadian version of the great flood described in the Bible. In the latter part of the century German architect Robert Johann Koldeway (1855–1925) spent twenty-five years carrying out depth excavations in Babylon and developing new archaeological techniques.

Perhaps the most romantic of those early avocational archaeologists was Heinrich Schliemann. Schliemann was a German businessman active in trade with Russia and America. He had always been interested in ancient Greece, especially the Homeric epics. At his own expense he began excavations in Ithaca in 1868, and in 1871 while excavating at Hissarlik in Turkey he discovered Troy, thus confirming by means of archaeological investigation the famous Homeric account of the Trojan War. Throughout the 1870s, Schliemann continued his work in Troad and later Myceniae, further demonstrating the ability of archaeology to provide deeper insight into the past, and thereby opening up a fresh chapter in archeology.

Other revelations followed; Flinders Petrie's excavations in Egypt; Arthur Evans's discovery of Minoan Crete; the work of Sir Leonard Woolley at Ur; the discovery of the Hittite, Sumerian, and Indus civilizations; and the discovery by Howard Carter and Lord Carnarvon of the tomb of King Tutankhamen with its wealth of information on the civilization of ancient Egypt. This expedition in particular reveals the thrill of discovery and the wonder of ancient times which was felt by scholars and by the public alike.

THE EMERGENCE OF ARCHAEOLOGY IN THE SCIENTIFIC MODE

Interest in the past as seen through artifacts had long been the domain of antiquarians. The road from antiquarianism to modern archaeology, however, began in the eighteenth century when some collectors, in tune with the spirit of the times, became more interested in understanding history through artifacts from the past than in merely collecting antiques and curiosities. Sir Richard Colt Hoare, for example, said "we speak from facts, not theory," and art historian Johann Joachim Winkelmann, who greatly influenced the neoclassical movement, placed art history on a more systematic basis than before. Winkelmann was also an archaeologist more in the modern than in the antiquarian sense, for he was the first to apply in the study of history on a comprehensive systematic basis, categories of style based on empirical observation, in his case the study of the Greco-Roman period. For this reason he has often been called one of the founders of scientific archaeology. Toward the middle of the nineteenth century archaeological excavation, classification and analytic techniques become more scientific.

Augustus Pitt-Rivers (1827–1900) was an early figure in both ethnology and archeology who pursued his studies within the theoretical framework of evolution. He was born Augustus Henry Lane-Fox but was required to change his surname when he inherited the estates of his cousin, the sixth Baron Rivers. Pitt-Rivers was a soldier by profession but devoted his life to studying cultural variation around the world and over time, assembling thousands of artifacts, 22,000 of which are now at the Pitt-Rivers Museum at the University in Oxford. His estates also provided fertile ground for excavating the Roman and the Anglo-Saxon past of England. Pitt-Rivers employed the careful systematic methods of science in his investigations, thus influencing the work of later archaeologists. For that reason he is widely consider the father of scientific archaeology in Britain.

In 1991, Colin Renfrew and Paul Bahn wrote that the surveying and excavations of archaeological sites have several features in common with military operations. Complex archaeological field projects, they say, are seen in terms of "objectives" and "campaigns"—both of which are military terms—which often involve the recruitment, funding, and coordination of large numbers of people. It is no wonder, they say, that Pitt-Rivers and Mortimer Wheeler, another early pioneer of scientific archaeology, had military backgrounds (Renfrew and Bahn 1991: 61).

Another difference between Pitt-Rivers and the antiquarians of his day was his interest not only in objects of artistic value, but also ordinary objects that gave insight into the life of ancient peoples. Pitt-Rivers arranged his collections according to type of object, and chronologically within each type, to illustrate artifacts in their evolutionary line of progression. This approach to collections was an innovation in museum display, and thus another of Pitt-River's contribution to science and its public interpretation.

Another pioneer in scientific archaeology was Flinders Petrie (1853–1942) who held the first chair of Egyptology in Great Britain and who developed a dating system based on pottery and ceramic fragments. By the first decades of the twentieth century the prehistoric and historical past were being thoroughly explored on a sound scientific basis not only in Europe and the Middle East, but in Asia and the Americas as well.

ETHNOLOGY AND THE THEORY OF MAN

Over the next two centuries information accumulated about distant and exotic peoples around the world, and with it a popular fascination for what had been discovered and described. The Age of Discovery of the sixteenth and seventeenth centuries began the most extensive contact which Europeans had ever had with people of different societies and cultures. The long and increasingly more intimate contact which resulted produced a wealth of documentation on different customs, social types, and languages in the form of published travel accounts, diaries, and government and missionary reports, presenting a picture of the vast diversity of peoples on the earth, some of which correspond in their level of cultural development to the prehistoric peoples discovered in archaeological excavations in Europe.

Thus by the later part of the nineteenth century a body of ethnographic data had accumulated which invited the same kinds of scientific classification as did the plant and animal world, and what that diversity meant for the overall history of the species. Some classifications were based on physical appearance such as skin color and the measurement of cranial size, while others were based on language and culture. This also raised the question as to origins of that variation and what it meant for the natural history of the species and its significance for the present. There were two different approaches to the origin of human variation; one held that there was a single origin of mankind despite the variation, and the other that the human varieties had different origins.

John Cowls Prichard was an advocate of the first of those positions. He was a physician whose medical dissertation in Edinburg in 1808 marked the starting point of a long career which ended in 1848 with the publication of the sum of his work titled *Researches* (cited above). The dominant biological science in those days was comparative anatomy, a field with which Pritchard was familiar as a physician. Comparative philology (known today as comparative or historical linguistics) also stood high in the first half of the nineteenth century since it had demonstrated, through systematic comparisons of different languages, genetic affinities whose origins lay in the ancient past, thus a similarity in method and in apparent form and process.

There was a strong tendency in comparative anatomy at the time to designate the non-white preliterate peoples of the world as a separate species of mankind. Pritchard, however, argued for the unity of all varieties within a single species, basing his claim on the biblical account of creation and the divergence that occurred after the Tower of Babel. He saw the "ethnological problem," as he put it, as a way of demonstrating that basic unity with reference to the findings and the methods of both ethnology and archeology; in his words "to trace the history of the tribes and races of men from the most remote periods which are within reach of our investigation, to discover their mutual relations, and to arrive at conclusions, either certain or probable, as to the affinity or diversity of origin" (Stocking 1987: 52).

His approach was therefore inclusive, with the attempt to establish connections between the races on the basis of similarities of physical type, religion, political institutions, customs, and above all language, a task which involved philology, archaeology, history, physical geography, anatomy, zoology, and physiology, but in which language and culture overshadowed physical anthropology (Stocking 1987: 51–52). Pritchard's work was continued by Robert Gordon Latham who had also studied medicine as well as philology. In 1850 Latham published *Natural History of the Varieties of Man* and in 1859, a two-volume work titled *Descriptive Ethnology*.

Interest in prehistory and ethnology grew in the first decades of the nineteenth century. In 1843 both the British Archaeological Society and the Ethnological Society were founded, and in 1848 the British Association for the Advancement of Science included ethnology within the natural history section. Another society was founded in 1832 as a result of a split as to the question of race. James Hunt (1883–1869) was a strong advocate of polygenesis and a vocal proponent of the superiority of the white race. The founders of the Ethnological Society, on the other hand, were advocates of racial equality. This belief was based on a Christian point of view and was part of their interest in evangelical missions.

In 1832, Hunt broke with the Ethnological Society and, together with the flamboyant Richard Burton of Africa fame, founded another organization which he called the Anthropological Society of London. The organization subsumed ethnology into a wider field of scientific investigation which he described as "the science of the whole nature of man." Ethnology, he said, was merely "the history or science of races," a science which he said was hamstrung by biblical dogma and distracted with questions of unity or plurality of the species. Hunt's purpose was to place the study of man within nature, in relationship to animals and the physical environment following the methods of science, with ethnology only a subdivision within that more inclusive science. The organization proved successful, especially since Darwinism had made the nature of Man of such widespread public interest. One unfriendly observer said that there was "nothing like it in any scientific body in the country" (Stocking 1987: 248).

EVOLUTION AND ANTHROPOLOGY

The idea of evolution was "in the air" when Darwin presented his version of the process and showed how it might work. The idea of progressive elaboration however, had its origins in the seventeenth century in

Descartes's *Discourse on Method* published in 1637, and was established in the new faith in Progress of the eighteenth century. According to that belief, mankind had advanced over time intellectually and morally as well as materially, and that it would continue to do so in the future. It was also believed that certain laws of nature were in operation and that careful observation and the exercise of reason would produce the same kinds of results as those elsewhere of the physical sciences.

With the emergence of Darwinian evolutionary thought at mid-century, says Glyn Daniel, "it seemed that prehistory was just proving the earlier stages of the long single pattern that led from the beginning of man through Victorian prosperity itself" (1962: 82), a time when contemporaries saw themselves at the pinnacle of a long process now traceable by means of archaeology to what the poet Alfred Lord Tennyson described as "the heir of all the ages in the foremost files of time." This evolutionary view also provided the conceptual framework for understanding the cultural variability which was observed around the world, from the simplest hunter-gatherer societies, to complex civilization and the intermediate forms in between.

The evolutionary paradigm in anthropology was later summed up by Malinowski (1944: 187) in his description of the cultural evolutionary theory of the time when he said that the evolutionary view and the comparative method—which was used with success in biology and linguistics—imply certain assumptions: (1) human beings are substantially similar, (2) they develop gradually from a primitive stage level and pass through various stages of evolution, (3) the common measure in their action and in their thoughts can be discovered by induction based on a vast survey of collated data, (4) the concept of survival is essential, for it serves as a key for understanding the continuity within transformation, and as a link between the various stages. What is strong and flourishing at one level becomes superstition at higher levels. Forms of marriage and kinship may ossify into terminology and survive merely as linguistic usage, long after the practice which engendered it has changed. And (5) as we move down the various developmental levels we find the most primitive stage accessible, "that is, the 'origins' of human institutions, customs and ideas."

A prominent figure in the development of that approach was an American, Lewis Henry Morgan (1818–1881). In his 1877 *Ancient Society*, Morgan made extensive use of Greek and Roman sources as well as information from the contemporary nineteenth-century world and his own extensive field work among the Iroquois. In that work he argues that the rise from savagery through barbarism to civilization can be accounted for in the interaction between the family, technology, and property within the larger social structure and forms of governance, along with intellectual forms. His theory received wide attention in Europe and influenced others in their attempts to understand the archaeological record and the different social and cultural forms encountered around the world. Among them were Karl Marx and Friedrich Engels. Marx incorporated Morgan's scheme into his theory, a work continued by Engels which appeared in 1884 in his *The Origin of the Family, Private Property and the States*.

The most prominent figure in Britain during that phase of anthropology was Edward Barnett Tylor (1832–1917) who appeared at the time when science was affecting every field, and when the evolutionary view had attained a high status. In 1861 Tylor published a book on his recent trip to Mexico, *Anahuac: or Mexico and the Mexicans,* and the following year he began attending meetings of the Ethnological Society where people of both archaeological and ethnological interests met. He was struck by the parallels between prehistoric tools found in Europe and those of contemporary non-literate people. He interpreted those parallels in Darwinian terms and came up with a theory of cultural evolution which proved quite popular and compelling to the mind of that period. In 1865 he published *Researches into the Early History of Mankind and the Development of Civilization*, and in 1871 he presented his theory in a book whose title revealed his broad approach, *Primitive Mind: Researches into the Development of Mythology, Philosophy, Religion, Language, Art and Customs*. Ten years later he published a popular textbook, *Anthropology: An Introduction to the Study of Man and Civilization*. Tylor was highly regarded by scholars of all disciplines, and his writings had a wide appeal.

Tylor had no field experience with the exception of his trip to Mexico and in 1888 a visit to the Pueblo Indians. His data came from the then available record which was of uneven quality. However Robert Lowie says that he was no armchair anthropologist. He gives examples of Tyler's observations in England which indicates his view towards data. Once he received a Tasmanian skin-scraper which he took to the butcher to have it tested. He also peered in shop windows for parallels of the Oceanian pump-drill (Lowie 1937: 69).

Tyler's strength, when it came to data, was in his critical sense in evaluating his source, and how he assembled facts into categories many of which became basic terms of current anthropology. Lowie says that Tylor had "an uncanny sense of fitness" in dealing with source material, providing a valuable service "in separating the dross from the gold from early chronicles and thus rescuing a substantial body of authentic fact on every period and phase of civilization" thereby providing scholars with a solid core of knowledge (Lowie 1937: 70–71).

An example of the kind of ethnographic categories he extracted from the data is that of cooking meat in wooden troughs filled with water, which is kept boiling by throwing heated stones into it. He brought that custom together with similar North American and Polynesian practices to create the ethnological category of "stone-boiling" and to define it as a stage in culinary development (Lowie 1937: 15). He also "blocked out the essential processes of primitive fire-making" from the written sources. And from a systematic comparison of descriptions of beliefs in soul and spirit he coined the term "animism" which he used to describe his minimum definition of religion. He also introduced other terms which have become standard parts of the anthropological lexicon. The best known of his definition is that of culture which appeared in his principal work *The Primitive Mind*, a definition which is essentially the same as the one used today.

Tylor's interest and his synthesis of data were holistic since rather than focusing on one system, for example religion and mythology, he embraced the whole field of ethnology and kept abreast of prehistory. He also showed a continued interest in linguistics. His approach was also self-consciously scientific, prompting him to offer proof based on a careful assessment and theoretically motivated organization of the available data. He also substituted the antiquarian's interest for an attempt at explaining facts, and rejecting a metaphysical concept of cause for the mathematical concept of function (Lowie 1937: 80–81). Tylor's theory of unilinear evolution however did not hold up, but his careful scientific approach set a stamp on the discipline he helped develop.

James Frazer (1854–1941) was another major player in the anthropology of that time. He was a classics scholar whose major interest was myth and religion. After reading Tylor's *Primitive Culture* he turned to ethnography for further data. His master work was *The Golden Bough* which appeared from 1906–1915 in twelve volumes in which he combined historical material from Classical Civilizations with ethnographic material from the published literature, as well as information from correspondence with missionaries and colonial officials around the world. He never went to the field and when once asked if he had ever seen a savage he supposedly answered "Heaven forbid." Yet his work proved highly influential for the way he wove the diverse material into a coherent theory. His unilinear evolutionary three stage theory of Magic to religion to science, soon proved untenable, yet the wealth of facts he gathered still provides a valuable source of information. He was also the first to describe the relationship between myth and ritual, and his inferences on the magico-religious way of seeing and interacting with the world is useful. His work was not only popular but it also inspired and influenced later field anthropologists.

Frazer has been criticized for the unilinear evolutionist theoretic framework in which he evaluated evidence and in which he argued his positions, yet his range of knowledge was broad, from classical and biblical sources to the ethnographic record then available. Bronislaw Malinowski, who was the role model for field work and advocate of a different theory, said, "Frazer's artistry, as well as his scientific sense in achieving a real synthesis from scattered and unrelated ethnographic evidence, appears at its best in his descriptive volumes on *Totemism and Exogamy*"

(1944: 191). His treatment of facts, wrote Malinowski, "is supremely contextual: the various aspects of human culture and human concerns became interrelated; the occasional comments and the collateral evidence is illuminated with real insight into human motives" (1944: 189).

British anthropologists not only sought information from people in the field abroad, but read the work of their colleagues on the continent and exchanged ideas with them. There were differences in their approaches, since those scholars were coming from different intellectual traditions tempered by conditions in their respective countries, yet overall three areas of general agreement emerged among them: that mankind formed a single species with a common underlying unity, that culture progressed by stages to the present time, and that the scientific method was the appropriate way of studying the physical, cultural, and linguistic variation of the human species.

THREE

One Discipline, Three Ways

The other two lines of anthropological research in Europe during this period, the French and the German, developed under roughly the same conditions, taking different approaches but ultimately contributing to what would become an international anthropology with different variations.

THE CONCEPT OF CULTURE

The concept that most defines the field of anthropology is that of culture. The term was developed within the German tradition and was defined by Gustav Klemm as "customs, information, and skills, domestic and public life, in peace and war, religion, science and art" and in 1871 by Tylor as "that complex whole which includes knowledge, belief, art, morals, law custom and other capabilities and habits acquired by man as a member of society." A generally accepted operational definition today is "The sum total of what an individual acquires from his society—those beliefs, customs, artistic norms, food habits, and crafts which come to him not by his own creative activity but as a legacy from the past, conveyed by formal or informal education" (Lowie 1937: 3). Culture in this sense is one of the defining features of the species and the focus of an entire branch of anthropology, a product of the confluence of the English and the German intellectual traditions.

ANTHROLOGY IN THE GERMAN TRADITION

Both folklore and Oriental philology formed an important part of nineteenth-century German scholarship. The leading figures were Gustav

Klemm (1802–1867), Theodor Waitz (1821–1864), Johann Jakob Bachofen (1815–1887), and Adolph Bastien (1826–1905). There had been in Germany interest in the cultures of preliterate people, which Herder had called *Naturvölker* ("natural people" as opposed to those of civilized countries). The study of such people was called *Völkerkunde* (ethnology) as opposed to *Volkskunde* (folklore).

Klemm conceptualized history as a unified process which progressed through three stages, periods of savagery, tameness, and freedom, a vision which included non-European peoples in his broad picture of human history. Klemm was also a collector of ethnographic objects and included a substantial amount of ethnological data in his ten volume *Allgemeine Culturgeschichte der Menscheit* (The General Culture History of Mankind) published between 1843 and 1852. Tylor praised his work as "an invaluable collection of facts," and a later commentator says that Klemm "laid out the first visionary place of systematic ethnographic collecting of museum displays" (Gingrich 2005: 79).

Waitz was attracted to anthropology through psychology, which was his specialty. He collected a massive amount of data on native peoples of the world which he published in six volumes titled *Die Anthropologie der Naturvölker*. He was also the first within the German anthropological tradition to proclaim the theory of monogenesis, arguing that the outward appearance of the varied people of the world was the result of different factors such as climate, social behavior, and cultural practice. The first volume of Waitz's work was translated into English and appeared in 1863 under the title *Introduction into Anthropology*, and was selected by the London Anthropological Society as an outstanding work from the continent. It was Waitz's arguments that further convinced Tylor that racial factors were not decisive in social and cultural variation (Gingrich 2005: 80).

Both Waitz and Klemm provided a great amount of ethnological data to scholars of the time. They differed, however, on the issue of origins, with Klemm arguing polygenesis and emphasizing material conditions in the process of cultural development, while Waitz argued for monogenesis and emphasized mental variety and progress.

Adolph Bastian (1826–1905) was a physician who traveled extensively and worked for a time as a ship's doctor. Like Waitz he supported a monogenetic theory but focused more specially on *Völkerkunde* with special interest in detailed case studies and wide comparisons, but also connections between cultures. He also emphasized description and local terminology. His work was within the tradition of Herder, Alexander von Humboldt, and Wilhelm von Humboldt. Bastian was also acquainted with the work of Wilhelm Wundt (1832–1920) a physician, physiologist, and philosopher at the University of Leipzig, who is acknowledged as the founder of psychology in general and of experimental psychology in particular.

He is most widely known for the psychic unity of man, the proposition that all mankind despite racial and cultural variation shares a common, basic framework of thought. This point had also been made in England by Pritchard and Tylor, in fact psychic unity had been one of the main stays of Tylor's theory. Bastian's main contribution was the theory of *Elementargedanken*, a set of specific elemental forms of thought which operate in the same way despite racial and cultural variation. He called the overt manifestation of those basic underlying ideas *Völkergedanken* or "folk ideas," a collective entity in which the individual mind is embedded, each entity determined by different historical and geographical conditions. In this regard he said that the people of different geographical regions, which he called *Kulturkreis* would exhibit different traits. Bastian thus tied culture to its underlying cognitive base within the world of social interaction and interaction with the natural environment.

Rudolph Virchow (1821–1902) was another contributor to nineteenth-century German anthropology. Virchow was principally a physician who contributed to a number of fields of medicine. He was known as the father of pathology for his work in helping to discredit the humorial theory, but was most widely known for his work in cell biology. He was the first to identify leukemia cells. He is also known as the founder of comparative pathology for his development of a standard method of the autopsy procedure. In anatomy he, simultaneously, with Charles Emile Troisier, identified an enlarged left supraclavical node as one of the earliest signs of gastrointestinal cancer. He also coined the terms "embolism" and "thrombosis."

In 1869, Virchow founded the *Gesellschaft für Anthropologie, Ethnologie und Urgeschichte*, The Society for Anthropology, Ethnology and Prehistory (*Anthropologie* refers here to physical anthropology). That organization was influential in encouraging German archaeological research. He also made several fields trips himself in the interest of archaeology sometimes together with Heinrich Schliemann, the discoverer of the site of ancient Troy, to Asia Minor, Egypt, Nubia, and the Caucasus. In 1879, he visited the site of Troy and published *Beiträge zur Landeskunde in Troas* (Contributions to the Study of the Landscape of Troy) and in 1882 *Alttrojanische Gräber und Schädel* (Ancient Trojan Graves and Skulls).

Adolph Bastian for ethnology and Rudolph Virchow for physical anthropology provided guidance for anthropology in the late nineteenth-century German tradition, a tradition in which Franz Boas and Robert Lowie were trained and which Boas passed on to the first generation of anthropologists in what would become the American tradition of anthropology. Both Bastian and Virchow were physicians trained in the physical sciences, bringing that orientation to the study of the physical and the cultural variety of the species. As Gingrich describes it they both "postulated the unity of man in an empiricist manner." Within this framework *Völkerkunde* now comprised an ethnology that classified and generalized the strictly descriptive data ethnology which led to the kind of approach we see today in anthropology (Gingrich 2005: 86–87).

ANTHROPOLOGY IN THE FRENCH TRADITION

The leading work in France was that of Emile Durkheim (1858–1917) who developed a sound system of sociology focused on social relationships, along with an empirical method of describing and explaining them, thus building the foundation for modern sociology. He took culture into account in the form of what he called collective representations which he regarded as a part of the functioning social system. His earlier work was sociological in nature, the *Division of Labor* (1893), *Suicide* (1894), *The Rules of Sociological Method* (1895) and "On the Normality of Crime" published in *L'Année Sociologique* in 1895.

He turned to the study of primitive societies later in his career. In 1903 he co-authored with his nephew and successor Marcel Mauss, *Primitive Classification* (1903), in which he and Mauss argued that society was the source of human thought, which, like language and logic, were influenced by the environment in which they grow. They argue that group mythology and native systems of classification are derived from social categories. In 1915 Durkheim argued in his famous *The Elementary Forms of Religious Life* that religion is a social phenomenon through which emotional security and social solidarity are attained. His data came from rain dances among the Pueblo Indians, the Australian aborigines, and elsewhere. Durkheim's structural function approach influenced later British social anthropology.

Marcel Mauss (1872–1950) wrote on magic and religion in *An Essay on the Nature and Function of Sacrifice* (1898) and *Outline of a General Theory of Magic* (1902) together with Henri Hubert (1872–1927), an archaeologist and sociologist of comparative religion. Mauss's best known and highly influential book *The Gift* (1924) deals with the function of gift exchanges and patterns of reciprocity using ethnographic data from around the world.

Another figure at the time whose ideas were influential was Lucien-Levy Bruhl (1857–1939) who developed the theory that the primitive mind operates differently than that of modern man. The primitive mind, he wrote, makes no distinction between the supernatural and the real but uses "mystical participation" to manipulate the world around them. This mytho-poetic mode of thinking thus contrasts with that of modern man which is speculative and logical. Those ideas were expressed in a series of publications from 1910 to 1927, the best known of which is *Primitive Mentality* published in 1922.

Levy-Bruhl has been criticized for making too sharp a distinction between the two modes of thought, pointing out that primitives clearly think as logically about their activities as do modern people. We read in his posthumously published notebooks that he later revised his thoughts on the two ways of thinking. Yet in the distinction he drew he did indeed identify certain distinctions which today would be called different cognitive styles. And E. E. Evans-Pritchard, whose description of the Azande of Africa was derived from field work, credits Levy-Bruhl with some insights on the different world views of people whose belief in magic forms an integral part of their lives.

Those French thinkers, Durkheim, Mauss, and Levy-Bruhl, like Tylor and Frazer, advanced interesting theories and added to our understanding of other cultures, yet their data came not from direct field work but rather from the written sources of the time which varied in depth and quality. Mauss recognized this, as did the British theorists, and stressed field work in their teaching. Another figure who advocated and actively did field work abroad in the interwar period was Marcel Griaule (1898–1956), a student of Mauss, which he promoted as a scientific form of description and explanation.

The reason for this lag between theory and field work seems to lie in the French disposition toward the former rather than the latter. One exception was Arnold Van Gennep (1873–1957) who published a book on taboos and totem in Madagascar in 1904 and in 1906 myths and legends of Australia and in 1910 *La formation des legendes* from written sources. But he was also a real field worker, a researcher who has been described as the very antithesis of the predominately armchair anthropology. His field work which supported his theory, however, was not carried out in distant lands among exotic people, but rather mainly in France. Van Gennep was concerned with the role of symbols and the ritual process in individual and social life as well as with the living folk tradition. His best known work, which was of theoretical importance for ethnology, is *Rites of Passage* published in 1909. In that book he described transitions between social statuses in the course of one's life—birth, coming of age, marriage, death—each ceremonially marked with rituals. His work, largely ignored in France, was very much noticed in Britain, most prominently in the work of British social anthropologists.

At the same time, French sociologist Robert Hertz (1881–1915) had become interested in the role of symbols and ceremonies in society. His perspective was typical of French sociology at the time, that symbols in society are determined not by psychological but by social factors. Unlike other sociologists of the day, who viewed field work in Europe as coming too close to folklore studies, Hertz set out to seek data for his theory in ongoing communities. Later as an officer in World War I he became interested in social cohesion among soldiers and during the time he served, until he was killed in action in 1915, he collected songs and folklore from the troops who served under him. His earlier work was the study of hand movements and of death. In 1907, two years before Van

Gennep's *Rites of Passage* appeared, Hertz published his study of death and the ceremonies associated with it. In 1911, he made a trip to the village of San Bessu in the Italian Alps north of Turin to observe people participating in a Catholic cult.

Despite French involvement in overseas territories, ethnographic field work did not appear in quality and quantity in France until after the Second World War. The focus of French interest was mainly theoretical and appeared somewhat later than in Britain. The greatest French contribution to international anthropology, until the appearance of Levy-Strauss, was the model of social organization and its role in human culture.

THE LINGUISTIC THEORY OF FERDINAND DE SAUSSURE

Another major figure within the French intellectual tradition of the time was Swiss linguist Ferdinand de Saussure (1857–1913). Saussure was trained in the tradition of comparative and historical linguistics. His dissertation dealt with a grammatical category in Sanskrit, and his university lectures covered topics such as Sanskrit, Gothic, and Old High German, yet his ideas on language in general were original and their impact on linguistics was enormous, an approach to language which later influenced the study of culture. His theory, however, did not take effect until after his death when in 1916 his students assembled and published his lecture notes under the title *A Course in Linguistics*.

Saussure introduced the concept of the arbitrariness of the sign; that is, the linguistic sign should be understood not in terms of essences but rather in terms of their relationship within the structure of language, a structure which consists of the grammar and sound system of language. He called that structure *langue* which he separated from *parole* which refers to the use of language in the form of speech. *Langue* showed regularities which persisted as structures over time and could thus be described in terms of patterns. Speech, on the other hand, varied with speaker, situation, individual ability, and so forth. The proper task of the linguistic, said Saussure, was the study of the structure of language, leaving the study of speech to others.

Besides the arbitrariness of the sign and the nature of linguistic structure, Saussure also introduced the idea of synchronic and diachronic levels of analysis. Synchronic descriptions describe the inner structure at any given point in time. Diachronic linguistics on the other hand is a study of how that structure changes over time. Saussure's model of language has proved highly productive and lies at the heart of contemporary linguistics. This view, together with Humboldt's concept of the inner form of language and its relationship to thought and culture, constitutes the heart of modern linguistic anthropology, a relationship which is also of increasing interest to linguists in general.

CONCLUSION

During the nineteenth century, the different but collateral, and sometimes intersecting, strains of inquiry (prehistory, ethnology, physical anthropology, and linguistics) developed in different ways and in different relationship to each other in Britain, France, and Germany, going in different directions in some cases, some of which were modified or abandoned by later generations, yet at the same time feeding into one another to produce three intertwined traditions that would result in a transnational disciple of anthropology seen in scientific and loosely defined holistic terms. Conditions in America created a fourth tradition which would be later reshaped mainly by the German tradition, but also by an important stream of British social anthropology with its Durkheimian strain, and later influenced by French anthropology.

FOUR

American Anthropology in the Nineteenth Century

The anthropological tradition in the United States developed as a result of the firsthand contact of people from European extraction with the native population; for a long time native North America was its focus. Two important contributors to the study of Native American Indians were James Adair (1709–1783) and John Heckewelder (1743–1823) both of whom tried to understand the native people and to describe them from what we see today as a modern anthropological point of view.

James Adair was born in County Antrim, Northern Ireland, and began trading with the Cherokees and the Catawba in 1735. In 1744 he moved his operations to the Chickasaw in what is now northern Mississippi. He spent forty years trading with, and living among, the Indians. During that time he learned their customs and their social arrangements from close observation and through participation in their activities, in one case as a member of a war party. In 1774 he published *The History of the American Indians: Particularly Those Nations Adjoining to the Mississippi, East and West Florida, Georgia, South Carolina and Virginia,* which includes the Catawba, Cherokee, Muskogee (Creek), Choctaw, and Chickasaw. In the preface he writes, "I sat down to draw the Indians on the spot—had them standing many years before me" and that he had "lived with them as a friend and brother."

Participant observation was not the only thing he has in common with modern ethnographers, another is to describe the people in their own terms. One of the reasons he gives for writing about the Indians was to dispel popular myths about them. "With irrepressible concern," he writes, "I read several imperfect and fabulous accounts of the Indians already given to the world." He goes on to assure his readers that "fiction and conjecture play no part in the following pages." He also writes about what we would call his field experience, describing the jealousy and secrecy of the people with whom he lived as well as their clannishness, an experience familiar to many in the field today.

Where Adair departs from contemporary anthropology is his theoretical position, for he devotes the first half of his book to the thesis that the Indians were descended from the lost tribe of Israel, a not uncommon opinion in those days, one that had been held by Spanish missionaries and by such prominent figures in the English colonies as Cotton Mather, John Elliot, Roger Williams, William Penn, and others. Otherwise Adair's ethnographic account bears a striking resemblance to modern anthropology as pointed out by historian Robert H. Berkhofer in his introduction to the 1968 reprint of the book. Adair's concept of culture, says Berkhofer, even though he doesn't call it that, is similar to the one held today, and the way he organized his account parallels that of John Swanton's 1946 authoritative summary of the Southeastern Indian tribes, a book for which Adair's account provides much ethnohistorical material.

Adair was not the only one in the English colonies who knew native peoples intimately. He was, however, the only one who described them in such detail. One of the most prominent of that number was Sir William Johnson (1715–1774), an Anglo-Irish trader who had lived among the Mohawks, had married into the tribe, and who had learned their language and customs. Johnson became the British agent for the Iroquois League (of which the Mohawks were members) and eventually the British Superintendant of Indian Affairs for the Northern Colonies. George Croghan (1718–1782) was another. Croghan was born in Ireland and became a fur trader in the Ohio Country where he also served for fifteen years (1752–1767) as deputy agent of Indian affairs under Johnson. Closer to home was George Galphin, born in Ireland in 1737 and later settling in the Georgia and the South Carolina backcountry to engage in the Indian trade. Scottish born Lachlan McGillivray (1718–1799) was yet another. McGillivray came to Georgia where he married a woman of the Creek nation. He was the father of Alexander McGillivray and the great uncle of William McIntosh and William Weatherford, three of the most notable chiefs of the Creek nation. Adair acknowledges Johnson, by then deceased, as a friend and goes on to dedicate his book to the others who he said had encouraged him to write it. It is significant that such individuals, who one might consider his peers, thought enough of Adair's knowledge of native culture to support him in that way. His book was read in Britain and America and in 1782 it was published in German translation. It remains a valuable source of information on the tribes of the Southeast at a pivotal time in history.

John Heckewelder was a Moravian missionary who worked for fifteen years among the Indians of Pennsylvania. In 1792, at the request of the Secretary of War he accompanied General Rufus Putnam to the Ohio Country where he helped in negotiations with the Indians at Post Vincennes on the Wabash River, and in 1793 he assisted in negotiating a treaty with the Great Lakes tribes. Heckewelder was also a member of the Philosophical Society of Philadelphia where he published articles in the society's transaction on native archaeology, ethnology, and linguistics.

In 1818, he published his principal work titled *Account of the History, Manners and Customs of the Indian Nations Who Once Inhabited Pennsylvania and the Neighboring States*. The book was translated into German in 1821 and the following year into French. Among his linguistic writings is a study of Indian place names in Pennsylvania, New Jersey, Maryland, and Virginia, published in 1822, as well as a comparative vocabulary of the Algonquin dialects which was taken from his manuscripts in the Philosophical Society and published in 1882.

As the nineteenth century progressed, two other figures appeared in the study of American Indian languages and cultures. One was Albert Gallatin (1761–1849) and the other was Henry Rowe Schoolcraft (1793–1864). Gallatin was born in Switzerland and immigrated to the United States where he had a varied and successful career. Among other things Gallatin founded New York City University and served as the United States Secretary of the Treasury. A major interest of his, however, was the study of American Indians and especially their languages. He befriended a Cherokee chief who supplied him with information on the structure and vocabulary of the language, and he was later engaged in research on the Pueblo societies as well as among the Pima and the Maricopa. In 1826, he wrote *A Table of Indian Languages of the United States* and in 1836 *Synopsis of the Tribes of North and South America*. Gallatin held the theory that the American natives had a common ancestor that came to the New World from Asia in prehistoric times, a speculation that would later be substantiated.

Henry Rowe Schoolcraft was a geographer, a geologist, and an ethnologist. In his youth he took part in an expedition to the headwaters of the Mississippi. He also served for a time as the U.S. Indian agent in Michigan. He married a woman of Scotch-Irish and Ojibwa descent from whom he learned the language and from whom he gathered information about the culture of the tribe. In 1846 he was commissioned to do a major study of the native population titled *Indian Tribes of the United States* which appeared in six volumes between 1851 and 1859. These early figures in American Indian studies represent the kind of work done by those individuals who had the opportunity, the interest, and the ability to describe Native Americans up close, contributing to the field before it became institutionalized.

EXPLORING THE AMERICAN WEST AND BEYOND

The beginnings of research within an institutional framework were in the form of expeditions sent by the government to explore the land to the West, thus providing both motive and opportunity for describing Indian languages and customs. The first of those expeditions, and by far the best known, was the Corps of Discovery, better known as the Lewis and Clark Expedition. The expedition was organized by Thomas Jefferson (1743–1826) with the purpose of exploring the vast region acquired in 1804 by the Louisiana Purchase. Meriwether Lewis (1774–1809) and William Clark (1770–1838) were chosen to head the expedition. The party left Saint Louis in May 1804 and proceeded up the Mississippi to the mouth of the Missouri, then up the Missouri to the continental divide and from there to the Columbia River and thence downstream to the Pacific Ocean, arriving on the coast in November 1805. They began the return trip in March of 1806 reaching Saint Louis in September of the same year.

The primary purpose of the expedition was to gather useful information on the newly acquired territory, establish trade with the native tribes, and to establish an American presence on the Pacific before the British could lay claim to the region. Another part of the expedition, however, was scientific, in keeping with the times and with the interest of President Jefferson. The ethnological part of the project consisted of recording the names and locations of tribes, collecting artifacts, and describing the customs and the social arrangements of the natives they encountered. Winter camps in close proximity with the natives meant that the expedition had time to take a somewhat more intimate look into native customs and languages than would have been the case with a direct traverse of the territory. Some of the tribes that were visited were the Crow, Mandan, Black Feet, Shoshone, and Sioux.

Other expeditions with ethnological components followed. From 1867 to 1872, Geologist Clarence King (1842–1901) led the Geological Expedition of the Fortieth Parallel (also called the Fortieth Parallel Survey) in eastern California and into Wyoming. The United States Exploring Expedition of 1838–1842, also known as the Wilkes Expedition, explored the Columbia River and the Pacific Northwest. Its purpose was both scientific and commercial. In the Oregon Territory the expedition collected ethnographic and demographic data among the local tribes as well as some two hundred artifacts. Also Horatio Hale prepared an extensive description of the Chinook Jargon, a native lingua franca used by the tribes in that part of the country.

Seaborne expeditions were also undertaken in the mid-nineteenth century. Among them was another Wilkes venture, known officially as the United States South Sea Exploring Expedition of 1836, which surveyed the Pacific and the South Sea, eventually circumnavigating the world, collecting hydraulic and other data along with biological specimens and artifacts. And in the North Pacific, there was the United States North Pacific Exploring Expedition of 1853–1856, which surveyed those parts of the North Pacific, the Bering Straits, and the China Seas frequented by American whalers and traders in the China trade. There were also other scientific expeditions to the Arctic, the Orient, the Near East, and the Southern Hemisphere.

In 1867, John Wesley Powell led an expedition to the Rocky Mountains and to the Green and Colorado River region. And in 1869 he and his team explored those two rivers, and for the first time passed through the Grand Canyon. In 1870 Henry Dana Washington (1832–1871) along with Nathaniel P. Livingstone and Lt. Gustavus Doan, explored northwestern Wyoming and the area that would become Yellowstone National Park. And the following year geologist Ferdinand Vandeveer Hayden, also with federal funding, further surveyed the Yellowstone country. Another government explorer was Lt. George Montague Wheeler who, in 1872, led the Expedition West of the 100th Meridian. Ethnology of the United States in the first half of the nineteenth century, therefore, was largely associated with government funded and sponsored surveys of the still little known western half of the continent.

ETHNOGRAPHY AND SOCIAL THEORY

Lewis Henry Morgan (1818–1881), who we encountered in the previous chapter, was not a government agent but rather a lawyer who worked for the railroad. His ethnological efforts were not directed toward the exploration of new land and the people in it, but rather toward a careful examination of a native people closer at hand, one which had maintained a long association with white society: the Iroquois of New York state. Morgan's activities were concerned not just with description but also with social theory. His theory, however, was not based on speculation and second-hand data, but on active field work among the Iroquois, work which has been described as groundbreaking. He is especially noted for his deeper understanding of the role of kinship in the social organization of preliterate peoples, and his systematic study of kinship terminology in order to better understand that basic level of social structure. In the 1850s Morgan looked further afield than the Iroquois, seeking information on kinship in various tribes elsewhere, through correspondence with scholars, missionaries, and U.S. and colonial agents around the world. In order to standardize the kinds of information he sought, Morgan designed a questionnaire which foreshadowed the *Notes and Queries* of later British ethnologists. In 1871 he published *Systems of Consanguinity and Affinity*, a subject that would become a central focus in anthropology.

ARCHAEOLOGY AND PHYSICAL ANTHROPOLOGY

For the curious at the end of the eighteenth and during the nineteenth centuries, the numerous earthworks east of the Mississippi presented a mystery. The prevailing theory was that the elaborate constructions were beyond the capacities of the ancestors of the native populations to build, and that their builders were some mysterious vanished race. Thomas Jefferson was one of the curious. He looked into the matter by digging a trench at a burial mound on his estate. This project has been called the first scientific archaeological investigation in America, for Jefferson applied the scientific approach in his inquiry.

He did this by testing ideas against the empirical evidence, data he uncovered. He examined the different layers uncovered in the trench to see how many of the bones were less well preserved in the lower than in the upper layers. From that he deduced that the site had been reused as a burial ground on many separate occasions. He said that more evidence was needed, but based on what he found he saw no reason why the ancestors of contemporary Indians could not have been the builders of the mound. What Jefferson displayed in his excavation was in many ways the foundation of modern scientific archaeology, namely, logical deduction from carefully derived evidence.

The questions of the mounds also interested a postmaster in Ohio by the name of Caleb Atwater (1778–1867) who surveyed some mounds near Circle City, Ohio, ancient earthworks that at the time were disappearing and which are now completely gone. In 1820 he published his findings in the first issue of the transactions of the newly formed American Antiquarian Society. Two other Ohioans interested in the mounds were newspaperman Efrain Squire and physician Edwin Davis who between 1845 and 1847 excavated over two hundred mounds and surveyed many earthworks. They prepared cross-sections and plans from their excavations and adapted a classification based on function in their interpretation of the data such as burial place, building platforms, effigies, fortifications for defense, and so forth. In 1848 they published their landmark book titled *Ancient Monuments of the Mississippi Valley*.

Another figure in early North American archaeology was Samuel Haven (1806–1881). Haven was the librarian of the American Antiquarian Society whose knowledge of the sources formed the basis for his synthesis *The Archeology of the United States* published in 1856 which is considered the foundation of modern archaeology in this country. Both Squire and Davis believed that the mounds were the products of some vanished people. Haven, however, disagreed, arguing that the mounds were built by the ancestors of contemporary Indians, who he said were of great antiquity indicated by cranial and other physical measurements, a method that was part of physical anthropology at that time.

William Henry Powell was also interested in archaeology. In his youth in the Middle West, he dug into mounds while he was learning about geology. When he became director of the Bureau of Ethnology he turned his attention to the question of the mound builders, and in 1881 he named Cyrus Thomas as head of the archaeological program at the bureau with the task of settling the question of the mound builders. Thomas worked on the project for seven years excavating thousands of mounds, eventually proving that the vanished race did not exist.

Another controversy which engaged archaeologists of the time was that of certain artifacts that had been collected in the United States which resembled the Old Stone Age tools of Europe. The Bureau of Ethnology sent William Henry Holmes (1846–1933) to investigate. Holmes started out as a geological illustrator and was focused on the scientific procedures and rigor. For five years he collected innumerable artifacts and subjected them to rigorous investigation, eventually proving that they were actually the refuse of Indian implement manufacture in recent times. Their superficial resemblance between the Old World artifacts had misled others into believing that they were products of the same era. By means of his systematic methods he also devised a survey classification of aboriginal pottery of the eastern United States, and he studied ruins in the Southwest and Mexico. Holmes did not believe that Native Americans had reached the New World in very ancient times, a proposition that was finally supported by evidence that began coming in the 1920s. Efrain Squire eventually became a diplomat serving in Central America and Peru where he examined Maya and Inca ruins.

Another major figure is Frederick W. Putnam (1839–1915), curator of Harvard University's Peabody Museum of Archeology and Ethnology from 1875–1908 from which he directed archaeological excavations in thirty seven states and abroad, among them in the Ohio Valley between 1880 and 1895, and surveys west of the 100th meridian from 1874 to 1909.

LINGUISTICS IN AMERICA

With the discovery of Sanskrit and the work of Franz Bopp, Jacob Grimm, and others, interest in language was becoming an increasing focus of scholarship both in Europe and in America. A major figure of the new linguistics in the United States was William Dwight Whitney (1827–1894). Whitney had studied in Germany under Bopp and others, returning to become, in 1849, professor of Sanskrit at Yale. Whitney made multiple contributions to the developing field of linguistics. Among them were *Language and the Study of Language* 1867, and *The Life and Growth of Language* 1874 which outlined a linguistic science that transcended the kind of comparative philology from which it had emerged. Whitney's books were translated into several languages, and although they seem incomplete today, due to a natural increase in knowledge, in 1933 Leonard Bloomfield said of them that Whitney's work still served "as an excellent introduction to language study" (Bloomfield 1933: 16).

Like so many others of the time Whitney operated across disciplines, borrowing from one branch of science in the development of another. His work in the physical sciences was an assistant to his brother Joshua, a geologist, whom he accompanied on an expedition to the Lake Superior region in 1849. It was the geologist's principle of uniformitarianism which guided Whitney's perception of the scientific basis of linguistics; the idea that geological phenomena past and present could be explained as a result of observing processes that have operated over in a uniform manner through time. His extension of that observation to diachronic linguistics was an important contribution to the developing science of language and part of the context in which another branch of linguistic science developed; the linguistic study of American Indian languages which became a part of American anthropology.

THE NATIVE LANGUAGES OF NORTH AMERICA

Whitney however was focused on Indo-European and not the nature of languages in general, thus his contribution to the study of Indian languages was only indirect. Within that context Albert Samuel Gatschet (1832–1907) had studied languages, art history, and theology at the Universities of Bern and Berlin. His favorite subjects were theology and the Greek language, and he patterned his work on that of Wilhelm von Humboldt. In 1867 Gatschet published a book on the etymology of place names in Europe, tracing them back to their Celtic, Latin, French, and German roots. In 1868, he immigrated to America and took up residence in New York City.

German botanist and student of language Oscar Loew had assembled sixteen vocabularies as a member of the Geological Survey West of the 100th meridian. He passed them on to Gatschet who had developed an interest in native languages. Some work had been done on Indian languages by Gallatin and Schoolcraft, but it was here that the study of the native languages of North America began, thus initiating a new branch of linguistic science. The task of this branch of linguistics was to collect, describe, and classify the languages of Native North America before they slipped into extinction, and in some cases to reconstruct language on the basis of philological work, work that went hand in hand with ethnological research, both carried out in a scientific manner.

Comparative studies appeared in annual reports of the Wheeler Survey in 1875 and 1876 and a year later in Germany under the title of *Zwölf Sprachen aus dem Südwesten Nordamerikas*, "Twelve Languages from the North American Southwest." Gatschet's work came to the attention of

Powell, then head of the Survey of the Rocky Mountain Region, who in 1877 offered him the position of ethnologist, which he accepted. He later became a member of the Bureau of Ethnology where he remained until his retirement in 1905.

Gatschet was active in the field working in 1877 with tribes in California and Oregon. He later published his study of the Klamath in monograph form. In 1881 he visited what was left of the Catawba in South Carolina, demonstrating the relationship of that eastern tribe with the Siouxian linguistic family of the western plains, and the likelihood that the eastern branch was probably the parent form. In 1881–1885, he studied the native languages in Oklahoma, Texas, and Southwest Louisiana identifying two new linguistic families and collecting associated ethnographic data. In 1886, he discovered and recorded the Biloxi and the Tunica languages, establishing the relationship with Siouan for one and identifying the other as representing a new linguistic family. Gatschet also extended his linguistic survey into Northern Mexico, recording some of the Carrizo language just before it became extinct, almost the sole survivor of the Pakawan family spoken in both Mexico and the United States. He also took advantage of frequent visits of American Indian delegations in Washington, DC, to elicit linguistic information. The work he began in recording Indian languages was continued by John Peabody Harrington who spent forty years recording endangered languages.

Besides active field work, Gatschet was also engaged in philological research, sifting through missionary catechisms and documents of other kinds to broaden and deepen our understanding of American Indian languages and their history. He studied the extinct Karankawa language of the southeast by using French and Spanish sources, and in 1881 he studied information collected from 1612–1614 by father Padre Pareja among the Taumean tribes of Northern Florida, showing that those extinct languages constituted its own linguistic family. In 1870 the Tutelo language of Virginia had been identified as belonging to the Siouan family, which in 1886 Gatschet confirmed by reconstructing the Biloxi of Southern Mississippi. He also published two volumes of ethnohistory titled *Creek Migration Legends* (1884 and 1885).

His range of interest and activity is seen in his membership in the American Folklore Society, the American Philosophical Association, the Washington Academy of Sciences, and the Anthropological Society of Washington and the National Geographic Society. The nature and scope within the field of anthropology and the community of scientists of his day represent the direction in which contemporary anthropology was heading in one of its four fields.

Another contributor to the study of Indian languages was Horatio Hale (1817–1896) who was the first to discover the affiliation of the Tutelo language of Virginia with the Siouan family of languages, and that Cherokee belonged to the Iroquoian family. He also wrote the *Iroquoian Book* of *Rites* based on interviews with tribal elders and interpretation from wampum belts. His linguistic studies, however, ranged beyond North America to include those of Polynesian, for he was commissioned to synthesize the data from the United States Exploring Expedition in the Pacific of 1838–1842 which appeared in 1846 as one of the volumes in the expedition's published reports, titled *Ethnology and Philology*. This publication is widely considered as the foundation for the ethnography of Polynesia.

Daniel Garrison Brinton (1837–1899) was trained and worked as a physician, but was also a prolific contributor to American anthropology, writing on native religion, folklore, and other ethnological topics. He was president of scholarly organizations of his time which encompassed the scientific nature and the subject matter of anthropology; the Numismatic Antiquarian Society, the American Philosophical Society, the American Folklore Society, and the American Association for the Advancement of Science.

One of the main characteristics of linguistic anthropology is its interest in the relationship of language and best known in the work of Edward Sapir and especially that of his student Benjamin Lee Whorf. That tradition, however, goes back to the work of David Garrison Brinton, his translation of an essay by Humboldt, and its application to the structural variety of the native languages of North America in his 1885 publication titled *The Philosophic Grammar of American Indian languages, as set forth by Wilhelm von Humboldt; with the translation of an unpublished memoir by him on the American verb*. In this respect Brinton was not only interested in grammar and phonology but also "the determination of the psychical character of the tribe through forms instinctively adopted for the expression of its thought and reciprocally the reaction exerted by those forms on the later intellectual growth of those who taught them as their only means of articulate expression" (quoted in Hymes 1983: 16).

FIELD ETHNOGRAPHY

The professional staff of the Bureau of American Ethnology included such field ethnographers as James Mooney (1834–1902) and Frank Hamilton Cushing (1857–1900) both of whom contributed to anthropology through their careful and empathetic field work. Mooney also contributed to linguistics with a provisional list of the principal tribal names and synonyms. His major contributions, however, were ethnohistorical research in the library and field work among the Cherokee, the Siouan Indians of the Southeast, the Cheyenne, the Kiowa, and the Sioux of the Dakotas. His reports were published by the Smithsonian. John Wesley Powell appointed him to the Bureau of American Ethnology in 1885.

Five years later the Ghost dance among the Sioux attracted public attention. This was only fourteen years since General Custer was defeated and there was concern in the government, fed by the press, that a new Indian outbreak might be brewing. Mooney was thus sent to "investigate this new, and threatening religion among the Sioux—and other affected tribes—and to report his findings" (Wallace 1965: v) and through his investigation he provided basic information on the broader cultural phenomenon of nativistic movements not only in North America but around the world. In this respect Anthony F. C. Wallace says that "it is remarkable, indeed, that so early a student was able to accomplish so much under such extraordinary difficulties (Wallace 1965: viii), a testament to his skill as a field worker. Mooney also saw the wider implications of what he had observed and recognized its relationship to ecstatic religions around the world. His ability to empathize with the Indians proved a great asset in his ethnographic work.

Another productive Bureau of Ethnologist field worker was Frank Cushing, who worked for many years among the Zuni. Once he had settled into the community he realized that he would have to learn the language in order to access the culture and its history. In time he came to recognize the richness of Zuni culture with its social complexity, its art, and body of folk knowledge. But he also realized that the culture contained a coherent system of belief and that that belief system could be revealed through the study of those aspects of Zuni life, but also in the songs and rituals of the priests who had preserved and distilled that view in the religious system. He also sought clues to those underlying beliefs in everyday activities. It became clear to him that to truly understand Zuni culture required time, patience, and sympathy.

In looking back on Cushing's work, Jesse Green says that "no other observer had been disciplined with anything like his scientific aims and methods, nor has participation in tribal life by any scientific observer after him extended so far or being central to the very mode of observation" barring some exceptions, "so admissible to the record" than that of Frank Cushing among the Zuni (Green 1979: 16). What Cushing did there, he says, "was nothing less than to develop a new methodology, participant observation," and that "by combing accurate observation with controlled imagination he was a century ahead of his time" (Green 1979: xii).

Cushing's contribution was in ethnography. His interest, however, also reached into the past with questions as to the origins and the history of the Zuni. To that end the privately funded Hemenway expedition of 1886–1888 was organized. The investigation included the systematic excavation of ruins, both those on the site of the Cibola towns and along the route that the Zuni migration took place according to their oral traditions. The expedition began in Tempe, Arizona, in 1887 with excavations which have been described as "probably the largest community in pre-Hispanic times" in Southern Arizona (Cushing 1979: 13). Failing health, though, forced him to withdraw from the expedition and the project was turned over to Jesse Fewkes who had been trained as a natural scientist specializing in marine biology.

Cushing has been described as a remarkable figure in anthropology spending much time living and interacting with the Zuni and thus practicing the method of participant observation which would become a mainstay of later ethnographic field work. He also advanced the idea that all people have culture which they draw on in their lives, and which can be systematically studied and described.

INSTITUTIONAL BASE OF AMERICAN ANTHROPOLOGY

Anthropology in roughly its four fields was carried on in the United States under the auspices or with the aid of several formal organizations. The earliest was the American Philosophical Society, founded by Benjamin Franklin in 1743, whose mission statement announces that it strives to advance knowledge in the broader definition of science which described in the words of its brochure the advancement and promotion of "scholarship in the humanities and social, mathematical, life, and physical sciences." In the early days the society fulfilled the functions of a national academy of science, a national archive and museum, and even a patent office. Two examples we have seen here is their help in the Lewis and Clark Expedition and in promoting and publishing the work of John Heckewelder.

Another important institution was the Smithsonian Institution, established in 1846 with original funding from a legacy left by British scientist James Smithson and administered by the U.S. government with the mission "increase knowledge." The museum housed specimens collected by various U.S. military and civilian expeditions, as well as artifacts of general historical interest, thus the nickname "The nation's attic."

The Bureau of Ethnology was established by an act of Congress in 1879. Its initial purpose was to transfer archives, records, and materials relating to Indians from the Department of the Interior to the Smithsonian, by which, in the words of its founder John Wesley Powell, was more broadly to "to organize anthropological research in America." Under his direction the bureau sponsored multiyear projects of archaeological, linguistic, and ethnological research with staff in ethnology such as Frank Cushing, Jesse Fewkes, and James Mooney and William C. Stutevant in linguistics, as well as support work by T. T. Waterman, Paul Radin, and Franz Boas. The bureau also collected artifacts for the Smithsonian, conducted exhibitions and expositions as well as acting as a repository for documents from various U.S. geological surveys. It also established a library with a collection of photographs and initiated a series of publications, most prominently its Annual Reports and Bulletins. In this way, it promoted what was at that time the still fledgling discipline of anthropology.

That fledgling discipline was also supported and promoted by museums in the latter part of the nineteenth century. One of them was the American Museum of Natural History in New York, founded in 1869, with the mission (in a broad Humboldtian sense) "to discover, interpret, disseminate—through scientific knowledge, research and education—knowledge about human culture, the natural world and the universe" in which Franz Boas served as an assistant curator of ethnology and somatology from 1896 to 1905. The Peabody Museum of Archeology and Ethnology at Harvard University was another whose director, Frederick W. Putnam, has been called the father of American archaeology.

CONCLUSION

Those loosely connected lines of pursuit, originally focused on Native America, were firmly bound together on a solid philosophical foundation in the early twentieth century by Franz Boas, and which now firmly established in the university as its institutional base, has spread its research out over the cultural of the world and has developed and has grown to the discipline we know today. To understand what that philosophical foundation is, and to see how it justifies the unity in a single discipline of such different fields as biology, linguistics, archaeology, and ethnology, we next turn to the question "what is science" and how it applies to the holistic scientific view of American anthropology.

FIVE

What is Science?

This brings us to the question of science. Can the designation science meaningfully describe a discipline so broad-ranging that it extends from biology to art? From the very beginning, anthropology was regarded as a scientific enterprise, most notably in the origins of academic anthropology in the United States at the end of the nineteenth century and the first part of the twentieth century. Yet many in the discipline today feel that sociocultural anthropology, unlike physical anthropology, has only a veneer of science, while others feel that it does not come up to the mark of a full science. Those perceptions, however, are not based on a sound philosophical consideration of what science is and how it has developed, but rather on the ordinary language meaning of the word which they take for granted. The best place to start when answering the question "what is science?" is with a semantic analysis of the word and the underlying conceptual category which it encodes, which have their own cultural history, the way ethnographers explicate the covert categories that form the basis of perceptions and behavior in any community.

The defining features of the English word "science" are "experimentation, replication, falsification, verification, prediction, use of mathematical expression, and the establishment of laws." Those features perfectly fit the laboratory sciences of physics and chemistry. Other sciences however do not correspond to all of those features, such as geology which is a field observational science where experimentation plays only a minor role, nor are astronomy and much of biology. But those disciplines deal with the material world where cause and effect, whether straightforward or in interactive feedback loops, is invariable under the same conditions and can be explained in terms of laws which make possible prediction and the formulation of exact knowledge which can be usefully applied.

This, however, is not the case for the human sphere where behavior depends to a great extent on the decisions of individuals and groups, decisions made within a system of meaning of their own creation, and thus creating more contingency and uncertainty in what is observed and, despite the application of much the same methods of systematic observation, classification, and rational inference within theories appropriate to the data, the same results are not achieved as those in the physical sciences. This has led to a division between the "hard" physical sciences and the "soft," some would say wanna-be social sciences. There is also a sharp demarcation between "science," both the physical and the social sciences, and the humanities, even when the humanities are pursued in the same systematic methodological way as are the other branches of scientific investigation. According to this definition the "science" of anthropology is only partially science and much of that falls within the realm of the soft social sciences.

This is because those fields of study operate with a paradigm which characterizes physical sciences in general, a paradigm which chemist Henry Bauer describes as a "voluntary consensus of *rational* opinion" about how science works based on its tried and proven success in explaining phenomena and in predicting outcomes.

Social science, to which sociocultural anthropology belongs, has no comparable consensus, and thus does not have the kind of coherent paradigm which forms the basis of the physical sciences; the physical sciences deal with predictable patterns of causality which govern the material realm, while social scientists deal with the human sphere with its many contingencies and thus its higher degree of uncertainty. This has led some adherents of the narrow folk definition of science to pronounce the social sciences as not real science at all, despite the name (Bauer 1992: 140). With this in mind we approach the question of anthropology as a holistic science in terms of two categories of inquiry, one that deals with the cause and effect patterns of the physical aspect of humans as a biological species, and the other that deals with the human sphere, both amenable to the scientific approach, but in different ways.

THE HISTORY OF THE WORLD AND THE CONCEPT OF "SCIENCE"

This concept of science, however, is not grounded in universal reality, but is the product of cultural processes with a clearly traceable history. In the case of English, this history shows a steady narrowing of scope as science emerged from philosophy (of which it was a part) to its status of an independent and a highly successful and prestigious intellectual and cultural system of its own. The modern word "science" is derived from the Latin *scientia* which, like the Greek word *episteme* (which inspired it) refers to any kind of knowledge that is organized in a systematic manner, and that is connected with some methodology and which is acquired by study. This was its meaning in the Middle Ages when the term was used to describe the seven liberal arts which constituted the basis for the medieval university curriculum: grammar, rhetoric, logic, arithmetic, geometry, astronomy, and music (Chantell 2002). It was in that sense that theology, due to its preeminence in the hierarchy of medieval learning, was sometimes described as the "queen of the sciences."

The Latin word *scientia* entered the English language in 1340 with the broad definition of "a branch of learning, skill" (Barnhart 1988). It was first recorded in 1725 with a meaning a little closer to its present sense, still retaining its breadth, as "a branch of learning based on observation and tested truths, arranged in an orderly system" (Barnhart 1988). Up into the first years of the nineteenth century "philosophy" was more commonly used for such studies. The word philosophy comes from the ancient Greek elements for "love of" and "wisdom." Wisdom, in its most general sense, means the "intellectual probing of an idea." Webster's Dictionary gives as the current meaning of "philosophy," "the rational investigation about existence, knowledge and ethics." Natural philosophy then was that branch of philosophy which dealt with the rational investigation of the physical world. Thus Dalton's 1808 study of chemistry was titled *New System of Chemical Philosophy*. Also Fourcroy's 1806 book on the same topic bears the title *Philosophie chimique*, and in 1806 the title of Lamark's book on zoology was *Philosophie zoologie*.

A new word was required as the ideal of a pure disinterested empirically testable science emerged from the broader field of philosophy. The term "science" thus came to replace "natural philosophy" in the general vocabulary. Science and philosophy are now separate intellectual endeavors with the emphasis in science on empirical observation and experimentation, while philosophy is left greater freedom of speculation, although the two are never really all that far apart, especially at the frontiers of empirical science. The earliest dictionary entry that gives a meaning of science in its current form is found in Murray's *New English Dictionary* of 1867. However, T. Merz, in his 1896 *History of European Thought in the Nineteenth Century* (vol I, p. 89), suggests that "science" had probably already acquired its current meaning as early as about the time of the formation of the British Association for the Advancement of Science in 1831, but was probably in use even earlier. In that respect we note that an influential polymath of the day, William Whewell (1794–1866), with a wide breadth of interest across the branches of knowledge, first introduced the term "scientist" in answer to a challenge from Coleridge, replacing "natural philosopher" or "man of science."

Once established as the designation of a new way of inquiring and knowing, the word underwent further changes, narrowing eventually to a concept which divides the physical sciences from the social sciences and which clearly demarcates both of those fields of study from the humanities (sometimes called the human sciences), which by general agreement forms a category in its own right.

THE DEVELOPMENT OF THE MODERN CONCEPT OF SCIENCE

The ideal of modern science was derived from the successes of investigators such as Robert Boyle and Isaac Newton who made the scientific study of the physical world stand out sharply among the other branches of learning and skill, a view that held that nature is governed by laws and that a phenomenon reveals regularity and orderliness which can be described in terms of laws that can be explicated and described by science. Throughout the Enlightenment and into the nineteenth century it was believed that the kind of universal laws discovered in the material world could also be found in the sphere of human thought and action, thus the retention of a broad meaning of science, yet one that found its semantic center on the presuppositions and methods of the physical sciences.

The question of how that concept could be extended to other branches of philosophy was first considered by David Hume (1711–1776). In the preface to his *Treatise of Human Nature* (1736), he outlines his view of a science that would encompass all of the formal branches of study, both those that had so successfully dealt with the material world—with objects that are outside of us—as well as those that dealt with the activities of humans, both theoretic and practical. In that latter category he included ethics, criticism, aesthetics, and political theory. His aim was to renovate those studies by introducing empirical methods, a task that could be achieved, he thought, with a method modeled on that of physics.

A century later, in 1843, another representative of the school of British empiricism, John Stuart Mill (1806–1873) published *System of Logic*, a work that was widely read in Britain and on the continent. In that work Mill took up the issues introduced by Hume, calling the study of human activities "the moral sciences." He shows how the empirical methods of the physical sciences can be applied to psychology and on that basis to an examination of the origins and nature of nations and racial types as well as to the study of "the actions to collective masses of mankind." He also says that more limited fields of study would appear in the moral sciences as they developed, such as political economy, but that those special fields would be subsumed under sociology (Hodges 1952: 18–20).

At the same time in France, August Comte was developing his own plan for a comprehensive "positive" science based on the scientific study of the physical world. His positivist philosophy appeared between 1830 and 1842 in his highly influential six volume work *Cours de philosophie positive*. Comte asserts that scientific knowledge, as defined by the physical sciences, was the only real knowledge, and that its application to problems of the economy and society would lead not only to a better, more scientific understanding of those human activities, but also to "progress and order," a phrase which became a slogan of the Positivist movement and is still emblazoned on the Brazilian flag. The science of society that he proposed he first called "social physics" and later sociology. It was from Comte's vision that the modern discipline of sociology developed. He hoped with his science-based philosophy to replace religion, a replacement which marked in his view the high point in the development of society. In that regard he viewed history as passing through three developmental phases from the theological, to the metaphysical, to the scientific. With this philosophy Comte made science a substitute for religion.

Robert Scharff in an article titled "Comte's Positivist Vision, Our Positivist Burden," describes Comte as the "poster child of scientific optimism" (2010: 435). His view of science, says Scharff, although no longer operative in its original ideological formulation, is by no means a relic of the past, but rather "serves quite well as a description of the background of most western technocrats as well as many Anglo-American epistemologists and philosophers of science and technology" (Scharff 2010: 436). This conceptual background has also been called the dogma associated with the physical sciences (Malinowski 1956). It is also the basis for the common conceptual or folk category, within the culture of both laymen and specialists alike.

In sum, the general concept of science that we know today, as laid down in the category labeled by the word "science," is not grounded in some universal invariable truth, but has been shaped by the tendency of natural categorization toward prototypical forms as well as by the empiricist and the positivist traditions of philosophy, by the pragmatic tendency that lends more weight to what seems to work the best, and by the simple force of custom.

COMTE'S VISION, OUR BURDEN

Whatever we call it—science or something else—doesn't really matter, as long as the job gets done in a spirit that emphasizes the advancement of knowledge through systematic methods appropriate to the field of study. But if the semantic-conceptual category interferes with that process, then it does matter, for at that point we encounter missed opportunities and misapplications leading to mistakes, and to distortions. The social sciences provide a number of examples of this tendency. Nobel Prize laureate, economist Friedrich von Hayek, criticized those who believed that the real world of economic activity could be measured with the same degree of precision as physical sciences. In his Nobel Lecture, titled "The Pretense of Knowledge," he said that economists did not have solid information for that kind of measurement. What they lacked, he said, could not be reduced to a number. Yet economists persisted in proceeding on the fiction that the factors which they could measure are the only ones that are relevant.

Niall Ferguson in his history of finance *The Ascent of Money* gives an example of what can go wrong when mathematically oriented economists focus too much on the mechanical-like aspects of behavior and leave out most of the reality they are dealing with. In 1997, Myron Scholes and John Merton received the Nobel Prize for Economic achievement for a model which was designed to price options in buying stock by by-passing guesswork for a more accurate assessment. They included five variables in their model which they reduced to a mathematical formula. It worked beautifully until it stopped working.

Long Term Capital Management, the company they had set up on those principles, worked well until in 1998, for no apparent reason equity markets dropped, resulting in volatility up and down and a loss of money. Next, following on a previous Asian collapse, Russia's financial system collapsed and stocks fell. This was not supposed to happen but it did and much money was lost. The effect on investors was what Ferguson describes as a "herd like stampede for the exit" so that everything went down at once. The designers of the elegant mathematical model had run up against something it was not designed to deal with, the human factor. Instead the model assumed perfectly rational actors who instantly absorb all new information and use it to maximize profits, not subjective irrational actors, "emotional human beings" who shift from greed to fear and sometimes act in a herd-like manner.

There are many other examples of the kind of distortions that occur in the social sciences when they strive too hard to accommodate to a model which works perfectly for one strip of reality but not for others, a tendency which has been called "physics envy." There is another way of looking at and applying science to the business of observation, description, understanding, and explanation, and that is the concept of science as it has evolved in German.

A BROAD DEFINITION OF SCIENCE

The German word for science is *Wissenschaft*, a word that belongs to the ordinary vocabulary of the language which is used on an everyday basis in the same inclusive manner by scientists as it is by non-scientists. The meaning of this word developed alongside that of its English counterpart in much the same way and within the same wider European intellectual tradition. In the nineteenth century, however, the two meanings went in different directions; the English meaning narrowing as seen above, while the German term after some contraction, regained the earlier broad definition, but now on a sounder philosophical basis. *Wissenschaft* is defined as the advancement of knowledge through

> The process of methodologically driven research, and the presentation of the results and the methods of research with the goal of transmitting specialized knowledge to teach scientific thought. Science begins with the collection, the ordering and the description of the material under investigation. Further steps are the development of the hypothesis and theories which must conform to the material through verification or replaced through falsification. The goals of science may be either pure or applied science and is traditionally divided into the fields of natural and humanistic sciences. (*Brockhaus Encyclopedia* 1965)

The usual division is *Naturwissenschaft,* the study of physical phenomena, the *Sozialwissenschaften,* which encompasses the social sciences, and the *Gesiteswissesnschaften,* the humanities. There are also various sub-divisions within each field, such as agricultural science (*Agrarwissenschft*), economics (*Wirtschaftswissenschaft*), the objective systematic study of law (*Rechtswissenschaft*), literary scholarship (*Literaturwissenschaft*), musicology (*Musikwissenschaft*), and so forth—in other words, any part of nature or of human endeavor is a branch of science if studied in the scientific manner, as defined above, and if it produces a systematic, theoretical body of knowledge on a given subject.

Erwin Schrödinger, one of the founders of quantum physics, made reference to this broad meaning of science when he addressed the question of the value of scientific research in a lecture delivered at Cambridge University in the early 1950s. To make a genuine contribution, he said, we must promote knowledge within a narrow domain by specializing. Yet science is more than utilitarian application for it has a value of its own across the range of inquiry. "I consider natural science to be very much on the same line as other kinds of learning—or *Wissenschaft,* to use the German expression—cultivated in universities and other centers for the advancement of knowledge" which he considers as consisting of "the study or research in history or language, philosophy, geography or history of music, painting, sculpture, architecture—or archeology or prehistory" (Schrödinger 2014).

THE HISTORY OF THE WORD *WISSENSCHAFT*

The word *Wissenschaft* is derived from the word for "to know" and appeared with the meaning of "a learned discipline" in the eighteenth century, replacing the word *Kunst* "art" which meant knowledge and skill of how to do something. In the early nineteenth century the word was used interchangeably with the word "philosophy," as it was in English and French. For example, Fichte's *Wissenschaftslehre* and Hegel's *Enzyklopädie der Philosophischen Wissesnschaften.* Both *Wissenschaft* and its English cognate "science," and their respective conceptual fields, narrowed in the early part of the nineteenth century (Makkreel and Lust, 2010).

About mid-century, however, this trend met resistance, especially in Germany.

PHILOSOPHICAL FOUNDATION

The first encounter was concerned with how history should be dealt with in a scientific manner. By then the study of history was acquiring a new consciousness of method and direction, while at the same time other disciplines were springing up "to support it and feed it with facts and principles" such as folklore, philology, and linguistics as well as ethnology, archaeology, and comparative mythology (Hodges 1969: 5–6). History from this perspective included the human sphere of nature which is different from the material sphere, and so the first attempts to answer the question of how science could be applied to human activity was concerned with the study of history.

The first move in that direction was taken by historian Johan Gustav von Droysen (1838–1906) in reaction to the positivist trend in the study of history which he thought was gaining ground. He expressed his views in the form of criticism of a book written by H. T. Buckle titled *History of Civilization in England*. The title of von Droysen's article was "Erhebung der Geschichte zum Rang einer Wissenschaft" (Raising History to the Rank of a Science). In his view there are two kinds of science (*Wissenschaft*), those sciences that deal with the physical world (*Natur*) and those that deal with the products of the mind (*Geist*). He only sketched out what those differences were. It was Wilhelm Dilthey (1833–1911) who more fully developed this idea.

Dilthey chose the term *Geisteswissenschsften* for the study of the human sphere from among alternative words, a term he made popular but one which first appeared in its currently recognized form as the German translation of Mill's "moral sciences." Dilthey was influenced by Mill's empirical approach, but rejected Mill's belief that the moral sciences could be usefully studied with the methods of the physical sciences. He saw no reason why one sphere of knowledge should not be as thoroughly studied as the other within this broader definition of science, but each in its own appropriate way. His "classic formulation" of the nature of the human sciences is in his 1883 *Einleitung in die Geisteswissenschaften*.

Dilthey regarded the human mind and nature as a unity, yet these two spheres of nature are divided between the physical world and the world of human thought and action. The study of each, he said, must be empirical and must avoid speculation. Also both are concerned with facts, yet facts of a different kind which thus require different methodologies. Both methodologies, however, still belong to the category "science" due to their empirical and systematic nature. The physical sciences seek abstracted explanation of universal validity. Data for the *Geisteswissenschaften*, however, are derived from the world of the human mind, of which the observer is himself a part. The observer's goal therefore is immediate insight into the data observed rather than a straightforward explication of cause and effect.

This aspect of the problem, *understanding* or empathy, became a part of the methodology that Dilthey worked out for the human sciences. Thus a duality arose as to the goals of science in general; *erklären*, explanation as in the physical sciences, and *verstehen*, understanding for the study of the human sphere of activity. Dilthey summarized his approach in three propositions: (1) individual facts (2) uniformities arrived at by abstract generalization, and (3) value judgments. The method of *verstehen* that he proposed has been described as "interpretive understanding," for to understand one must first be able to interpret the observed behavior. Dilthey's approach therefore expanded the old practice of hermeneutics to include the interpretation of not only written texts but also of spoken words and gestures and behavior in general, and through them an interpretation of the thoughts of individuals and communities.

Verstehen, as interpretive understanding, thus means getting into the minds of others. This is not as esoteric as it may seem on first glance, for that is what we do every day when we infer the intentions of others by observing what they do and say and how they do and say it. Dilthey's method of *verstehen*, therefore, is a more self-conscious and systematic elaboration of everyday inference in social interaction, in the same way that *erklären* is the systematic, self-conscious elaboration of practical reasoning. The two methodologies together, explanation and understanding, he argued, thus provide us with appropriate means of explaining and understanding the phenomenal world, each sphere in terms most appropriate to its nature.

Another aspect of the Dilthey's approach dealt with the question of the particular and the general. The physical sciences focus on particular events as parts of patterns or regularities. The understanding of the meaning and the value of the unique event, however, is also a valid goal of scientific study, for unlike material objects which are external to human actors, humans are not simply products of the push and pull of physical forces, for they create their own world by organizing their own meanings which often results in unique events or products that can also be studied in a systematic scientific manner.

While Dilthey was working out his philosophy of science, others were working on the same topic from the direction set earlier by Immanuel Kant. Kant's critical philosophy was aimed at explaining the nature and the limits of reason, but it focused mainly on the natural sciences, leaving still unexamined a more complete critical approach to the other branches of human knowledge. Various scholars later turned to Kant's work in an effort to bring it forward. Those scholars were known as neo-Kantians. They were divided into two schools of thought; the Marburg school concentrated in the University of Marburg, and the Baden or the Southwest school named for the universities in the part of Germany where its members worked. The founder of the Marburg school was Hermann Cohen (1842–1918) and its best known participant in this country was Ernst Cassierer (1874–1945).

The founder of the Baden school was Wilhelm Windelband (1848–1915) who argued that the subject matter criteria for dividing the sciences—physical world on the one hand and the sphere of human activity on the other—be dropped and replaced with the methodological distinction between the disciplines. One of those methodologies he called nomothetic since it establishes lawful uniformities. The other he called ideographic which describes unique historical events. We might add that the nomothetic physical sciences are applied to individual cases in order to understand events within the broader causal framework. We see this all the time when the laws of physics and chemistry are applied to understand, for example, why the steel structure of a bridge has failed or what killed a dolphin or seal found dead on the shore.

Dilthey's response was that not all physical sciences are nomothetic, for geology, biology, and astronomy are also ideographic, and that economics, linguistics, and aesthetics also arrive at generalizations. Windelband said that all that was true, but that his primary concern was the concrete operation of a system for which he felt his approach was more productive.

Heinrich Rickert (1863–1936) was a student of Windelband. He shared with him the anti-positivist goal of distinguishing the physical from the human sciences. Like Windelband (but unlike Dilthey) he was more concerned with method than content, and like Windelband he thought of the human sciences, and in particular history, as individualizing, involving reference to values, while the physical sciences were generalizing, dealing with universally valid laws. Furthermore he objected to the label *Geisteswisseschaften*, since the word *Geist*, "spirit," tended toward the supernatural or to the psyche. He therefore substituted the term *Kulturwissenschften* for *Geisteswissenschaften*.

In sum, despite their differences, Dilthey and the neo-Kantians of the Baden School all agree on the following points which form the basis for the present meaning of the German word for "science:" the division of systematic knowledge into the spheres which deal with the material world on the one hand and the world of human action and thought on the other; that both sciences must be empirical and not based on speculation; that there are methods appropriate to each, and that the human sciences should not rely on those of physical science as its sole model; and the necessity in the human sciences of subjective understanding. In this way they preserved and updated the holistic view of science, and the inclusive scope of the earlier concept whether labeled *scientia* or philosophy.

INTERPRETIVE UNDERSTANDING AND THE SOCIAL SCIENCES

The line of thinking developed by Dilthey and the neo-Kantians had a profound influence on German sociology in the first two decades of the twentieth century, as seen most prominently in the work of economist Werner Sombart (1863–1941) and sociologists Georg Simmel and Max Weber (1864–1920). Weber was a colleague and friend of Rickert during their days together at Freiburg. After the publication of Rickert's *The Limits of Concept Formation in the Natural Sciences* Weber wrote to his wife "I just finished Rickerts. He is *very* good" (introduction to Rickert 1986: vii). Also Rickert dedicated the third and fourth edition to the memory of Max Weber.

Weber brought the thinking of Dilthey and the neo-Kantians, with their view on *verstehen*, in closer relationship with one another as well as with the positive position which emphasizes causality, empirical verification, and generalization in terms of the search for laws. It was his work that most clearly applied those methods to the social science.

In regard to generalizations Weber says that some phenomena reveal laws of nature which are regular and orderly and which can be described in statistical terms and with other mathematical formulations, as well as with illustrative analogies. He cites Gresham's Law in economics as an example. But since man is both rational and emotional, to focus only on such regularities is to understand only part of what transpires on an average basis, for the emotional aspect of behavior can seriously affect the motivations which produce such patterns. Such examples show that the law-seeking task of the physical sciences is only a part of the task of accounting for behavior. This, however, is not a weakness of social science, but rather the inadequacy of an idealization of science which ignores the fact that reality consists of more than what an ideal version of science based on the physical sciences presents.

With that idea obviously in mind, Weber devised a social science that strives for both causal and meaningful adequacy which he called *Verstehende Soziologie*, or interpretive sociology. In biology, he says, we can comprehend (*erfassen*) the functions of a cell which can be studied in terms of laws, but we cannot understand it (*verstehen*), since a cell has no independent intension or motivation. When we turn to human action we can also find the kinds of regularities that can be observed and accessed in terms of probable outcome, and which can be statistically described. Yet also present in human interaction is the factor of meaning. Weber identifies two types of meaning. One is rational meaning and the other is intentional meaning. For example, we can understand the meaning of the statement two plus two equals four. But then comes the question of what ones means when one writes it out, or when an actor employs it in a business transaction.

This takes us to motivation, that is, to what reasons actors have for doing what they are observed doing. The motive of a wood-cutter for cutting wood, for example, is compensation that the work brings him. Weber calls this rational motivation. Another category is emotional motivation which involves anger, joy, enthusiasm, and so forth. Such emotions can be read from gestures, tone of voice, and other signs, and their link with certain behavior can provide the motivation for it. Therefore the external process of action and its simultaneous motivation are recognized as meaningful in their proper setting, and in this way can be understood. Thus meaning and causation are taken into account in Weber's interpretive sociology.

This is the way Weber envisioned a social science which involves both understanding and interpretation on one hand, and causal explanation on the other. There is, however, no general paradigm for such a combination across the range of the social sciences, similar to the more unified paradigm found in the physical sciences which is based simply on explanation. This diversity is one reason that some believe that the social sciences are not really sciences at all. But is such diversity really a defect? It all depends, says Outwaite, on what we want from the social sciences and the humanities. Is the goal a coherent body of laws, or a few theories isolated by well confirmed conditional propositions? Or do we want to be more concerned about deepening and systematizing and enriching with empirical data, our understanding of social life as we experience it as participants. Or do we simply want to know better how society works as best we can by means of the social science approach and as a better means of appreciation in the humanities? Or perhaps the goal is knowledge for the sake of prediction and control? Or perhaps we employ the methods as a way to orient social science to the political action within a group or a movement, or to better understand and predict trends in the market for commercial purposes. Another goal might be for the development of interpretive models for social behavior and for the better, more insightful appreciation in the humanities.

There is no single answer, but this is not a bad thing, for we might say that any way we can deal with the world of phenomenal experience by means of "the process of methodologically driven research with the goal of transmitting specialized knowledge," qualifies that enterprise as a scientific endeavor. This holds for the humanities as well as the social sciences, each with its own limitations and each with its own areas of greater understanding and its different possibilities of explanation (Outwaite 1975).

THE HUMAN SCIENCES (THE HUMANITIES)

In English, by general agreement, a sharp demarcation divides the humanities from the social and the physical sciences. This has been argued most prominently by C. P. Snow (a scientist and a novelist) who claims that the two spheres of activity involve two different cultures. That may be the case in some respects, yet the application of the scientific method, defined in this broad sense, gives the sciences and the humanities a kind of unity despite their diversity, since they both involve the reasoned contemplation of the phenomenal world. We take here as an example the way the study of literature is dealt with within the broad definition of scientific inquiry.

In English, literary scholarship is often described as literary criticism where criticism means the act of making careful judgments about the qualities of a literary work. This leaves out, or makes ambiguous, a lot of what literary scholars actually do, a range of activity covered by the German word *Literaturwissenschaft*. This includes the study, in a systemic manner, of genera, narrative, style, historical trends, etc. as well as individual authors and their individual works and their individual styles of writing in a systematic manner, much as the material world is studied by the physical sciences and human activity by the social sciences.

In an essay titled "Science and the Public Critic" (in his book *Anatomy of Criticism: Four Essays*), Northrop Frye says that literature "must be examined in terms of a conceptual framework derivable from an inductive survey of the literary field" (1957: 7). The word "inductive," he says, suggests some sort of scientific procedure. He then asks whether criticism is a science as well as an art. Not a "pure" or "exact" science, but one in which scientific principles are employed in the treatment of evidence. It is in fact that approach when investigating the past, he says, that separates history from legend.

Elsewhere he talks about the critical approach as one that describes the text and which studies its historical and cultural background, noting that seen from this perspective the relationship of literary scholarship approaches the area of the social sciences, a direction strongly resisted by many since it "cut across the conditioned reflexes of most humanists at the time" (1957: xviii). In other words this way of looking at literature violates the folk category distinction between the "humanities" and the "social sciences" despite the similarities.

Frye defines criticism as careful judgments about the good and the bad qualities of a literary work. "I realized early in my critical life," he said, "that evaluations was a minor subordinate function of the critical process, at best an accidental by product, which should never be allowed to take priority over scholarship." He goes on to say that "scholarship" may support one's judgment of a work of literature, but "scholarship will never be founded on the value judgment" (1982: xvi). It is here that he makes a distinction between the critic—in his words the "public critic"—and literary scholars who study texts, context and content in a more rigorous manner, basically the same distinction made in German between *Literaturwissenshaft* and the *Literaturkritiker* or *Feulltonist*, the latter term referring to those who write for the style and culture page of newspapers.

Frye goes on to discuss the distinction between the two activities and their value. In order to "appreciate" literature and to get in more direct contact with it, he says, we turn to the "public critic" whose task it is "to exemplify how a man of taste uses and evaluates literature, and thus shows how literature is to be absorbed into society." But here, he says, "we no longer have the sense of an impersonal body of knowledge" (1957: 8).

"The public critic tends to episodic forms like the lecture and the formulaic essay, and his work is not science, but another kind of literary art" (1957: 8). Nor does the public critic engage in theoretical constructions, but rather selects his ideas from a pragmatic study of literature. "There is no way of distinguishing what is genuine criticism and therefore assessing the progress made toward making the whole of literature intelligible or simple taste and therefore the vacillations of fashionable prejudice" (1957: 9). Frye says that the public critic "is apt to make only a random and haphazard use of his material." He compares such critics to Shakespeare's Hamlet who in the graveyard ignores everything else and picks up a skull and moralizes about it.

The thorough critic (*Literaturwissenschaftler*), however, is like a scientist, a fact that everyone who has seriously studied literature knows, for the mental processes involved in such study is "as coherent and progressive as the study of science" (1957: 10). The same kind of mental training is involved in both enterprises and in both "a similar sense of unity of subject is built up" (1957: 10–11). Physics is an organized body of knowledge about the physical world, he says, and not the thing it studies. A student will say that he is learning physics not nature, but through the coherent body of knowledge he will come to know nature much better. So it is with literature and literary scholarship. Frye's observations on the study of literature are applicable to the study of music, the graphic arts, architectural style, and the history of ideas, all topics grouped together as the humanities but that share common ground with other systematic, scientifically oriented areas of inquiry.

In dealing with the study of art in general, Thomas Munro has staked out a similar position when he talks about a new scientific approach to the humanities, one that is more objective "in the spirit of the empirical sciences without exaggerating the claims of scientific exactness or certainty," and "emphasizing more careful observation and descriptive analysis of the work of art as a kind of cultural phenomenon" (1967: ix). His goal, he says, is to help lay a theoretical foundation for this way of studying the arts. In these two examples (there are others) we see the recognition of the reality and the benefits of the broader view of science.

Aesthetic refers to the emotional value given to an object, or a form, or an action due to a quality that gives pleasure and satisfaction to the sense and to the mind, namely due to its beauty. The meaning of beauty differs from culture to culture, from age to age, and even among individuals, yet the concept and feeling are still there and expressed in different ways, but always in some manner expressed. Likewise, the aesthetic value differs, as do the ways of judging the aesthetic value of something, but, urge and the basic sensations involved seem to be universal. Appreciation is the recognition and enjoyment of what is considered the good qualities of something, and a full understanding and a more accurate perception of it and a true estimation and evaluation of its value. Both can be studied in the same systemic manner as are the context, style, and historical processes seen in works of art of all kinds.

To summarize, the elements which unite the physical sciences, the social sciences, and the humanities, in different degrees, are systematic observation, classifying, modeling, inference, trial and error, interpreting the patterns discerned, explaining them in causal terms or understanding how something works, and describing one's finding in the clearest and most appropriate manner, in expositive narrative, in appropriate analogies, and when all that proves inadequate and the data amenable, in the formal language of mathematics. Moreover the physical sciences and their purely theoretical subdivisions are not exclusively nomothetic, and the humanities not exclusively ideographic although that is the emphasis in the humanities, as nomothetic concerns are in the physical and the social sciences (Nagel 1961).

The emphasis which takes precedence depends on the subject matter and on what the application of the scientific method can yield. Mathematicians and physicists in their pursuit of knowledge enjoy a better fit with the ideal of science, which has led them to get up on their high horse, as philosopher Holm Tetens puts it; an unwarranted attitude, for physicists and mathematicians have "picked out the raisins from the cake of reality," those aspects of nature which can be measured and which can often be described with elegant mathematics, while others deal with those realms of reality where other factors are in play, and thus where methods, other than in those in the physical sciences are more appropriate (Tetens 2013: 78).

Given that perspective, the semantic range, and thus the shape of the concept science, is more accurately covered by the German word *Wissenschaft* then by the narrower range covered by the English word "science." Ramon Betanzos (translator of Wilhelm Dilthey) notes that this more restricted "English usage might well be expanded to advantage in adopting a more generic sense of 'science' that reflects German usage" (Dilthey

1988: 32). But that won't happen, for as David Lindberg points out, a community, once its meanings are fixed, is "unlikely to relinquish its favored usage without a fight" (1992: 2).

SIX

Converging Sciences

In the previous chapter we examined the history of the word and the concept "science" in two traditions, arguing that the broad definition of science is valid and rests on a solid historical and philosophical foundation. In chapter 7 we will see how American anthropology was a direct outgrowth of that broader view and thus deserves to be called scientific in all of its four sub-disciplines. In this chapter we turn to the question of holism: can a discipline that covers the entire range of science be a viable field of study? Our answer is not only that it can, but that science actually works that way, for as science burrows ever deeper into selected sections of reality disciplinary divisions increase, and with those divisions the sense of the unity which was a feature of nineteenth-century science began to fade. However as those separate disciplines reached out and became more interrelated, even interdependent, a counter trend of convergence set in whereby disciplinary boundaries began to blur. This was observed in 1967 in the biological sciences by Sir Peter Medawar, a process which, he says, reflects the tendency of modern science in general.

> One of the distinguishing marks of modern science is the disappearance of sectarian loyalties. Newly graduated biologists have wider sympathies today than they had in my day, just as ours were wider than our predecessors.' At the turn of the century an embryologist could still peer down a microscope into a little world of its own. Today he cannot hope to make head or tail of development unless he draws evidence from bacteriology, protozoology, and microbiology generally; he must know the gist of modern theories of protein synthesis and be pretty well up on genetics. (1967: 115)

In 1944, Erwin Schrödinger, one of the pioneers of quantum physics, was one of the first to suggest a connection between two important branches of the physical sciences, biology and quantum physics. Some of the insights and suggestions expressed in his book *What Is Life?* especially his concept of code scripting in reproduction and in growth, were read by genetic scientists studying DNA on their way eventually to describing the double helix and the genetic code (Cobb 2015). Only recently, however, has a greater convergence between biology and quantum physics begun in the form of the new and still tentative discipline of quantum biology.

Biology deals with molecules, cells, tissues, and organisms in the life process, while quantum physics focuses on the still more basic level of forces, atoms, and subatomic particles which constitute matter. Those two disciplines deal with different aspects of a single reality which are properly studied in separate disciplines. This new field however brings them closer together by examining quantum phenomena within life systems, thus breaking down a line of disciplinary demarcation within a single reality.

Some of the phenomena investigated are photosynthesis and cellular respiration in plants, cellular repair and animal vision as well as magnetoreception, or a sense that allows an animal to detect a magnetic field for perceiving altitude, direction, and location which might account for how migratory birds find their way over great distances. Such abilities have also been detected in bacteria, as well as in insects such as fruit flies and honey bees and in aquatic life such as lobsters, turtles, sting rays, and sharks. In these and other studies we see the possibility of an increase in understanding the physical world through the merger of two different scientific disciplines (Al-Khalili and McFadden 2015).

In the next two sections, we examine the natural process of convergence in two fields which are directly related to anthropology, and in which anthropology plays an active role; cognitive science and the disciplined study of language, both of which, like holistic anthropology, span the range from physical and social science to the humanities.

COGNITIVE SCIENCE

The comparative study of animal behavior has increasingly shown similarities in the way different species perceive and act upon the world and how they communicate with others of their own species. Yet despite that continuum there exists a cognitive gap between humans and all other species. No other species, not even those most closely related, comes anywhere near the level of cognitive ability of human beings, an ability seen for example in the way people imagine endless situations, the way they create scenarios of various kinds, relating them in narratives to themselves and to other people; stories that range from past to present experience, and that cross over into the world of imagination. Robert

Burns has put this species-difference poetically when he addresses a mouse that he has turned up in the field while plowing.

> But mousie, thou art no thy lane,
> Proving foresight may be in vain:
> The best-laid schemes o' mice and men
> Gang oft agley
> An' lea'e us naught but grief and pain
> For promised joy!
>
> Still thou art blest, compared wi' me!
> The present only toucheth thee:
> But och! I backward cast my e'e,
> On prospects drear!
> An' forward, though I canna see
> I guess an' fear!

It is that gap between the here and now, and what might be, or what might have been or what we can do to shape our own destiny in a conscious manner, that we find myth and religion which are traditional fields of anthropological inquiry and of philosophy and science. Consciousness was one of the phenomena that psychology attempted to describe and explain. In fact, William James, one of the founders of modern psychology, wrote that psychology should be "the description and explanation of states of consciousness." And another founder Wilhelm Wundt, along with his students, initiated laboratory experiments for more systematically studying mental operations. Yet during much of the twentieth century, the study of human cognition was ruled out by restrictions placed on the field by behaviorist psychologists. Their approach rejected any reference to consciousness, restricting the task of the psychologist to that of prediction and control of behavior in which introspection constitutes no part of the process. Their goal was to purge psychology of any "intangibles and unapproachables." Terms like "sensation," "perception," "image," "purpose," and even "thinking" and "emotion" were deemed "unscientific" and were excluded from their terminology since those words were subjectively defined.

In the 1970s this approach came under attack from linguist Noam Chomsky in response to the simple stimulus-response theory of language acquisition proposed by B. F. Skinner. Chomsky, in his *Language and Mind* fully agrees with the behaviorists' aversion to speculation in place of facts derived from careful observation and experimentation. "Obviously any rational person will favor rigorous analysis and careful argument," he writes. Yet the behaviorist scientists (with that phrase bracketed by quotation marks) "are merely mimicking the surface features of the natural sciences," concentrating on what Chomsky calls "peripheral issues." Moreover "there has been a natural and unfortunate tendency to 'extrapolate' from the thimble full of knowledge that has been attained" (Chomsky 1972: xi). Chomsky goes on to elaborate a model of language based on the cognitively based abilities to acquire and to use language.

Others pursued the study of behavior along similar lines, recognizing that humans "don't just run and mate and walk and eat," but that they also "think classify, analyze, and theorize" (*Stanford Encyclopedia*), recognizing that besides the outer behavior of humans there is an inner life, one that is in our heads where we are constantly active in the form of processing information, shaping it into meanings that map our experience, giving us a framework within which we can make sense out of our experience, knowledge on which we can make decisions as to how to act, and where we imagine other possibilities and strategies for action.

Cognitive psychology as now conceived deals with perception, reason, memory, motor skills, interpreting, imagining, problem solving, decision-making, the production and comprehension of language, and explaining. It is thus not only a defining feature of humans among other species, but it is also central to human action of all kinds, and is thus a topic of convergence from different disciplines bringing with them different bodies of accumulated knowledge, different theories, and different research strategies.

Cognitive science had its origins in the 1930s and 1940s in the early days of cybernetics with its implications for the organizational principles of the human mind. Another precursor for this branch of science was the work of Alan Turing and John von Neumann with digital computers and the theory of computation. The computer thus became a model for thinking about the mind. This led, in the 1970s and 1980s, to interest in artificial intelligence and attempts to simulate cognitive abilities with computers. In the 1970s chemist and physicist Christopher Languet-Higgins turned his attention to the brain and artificial intelligence and in 1973 coined the term "cognitive science."

Throughout the 1970s and 1980s neural networks and connectedness became a part of research paradigms, and the mind was envisioned as a set of complex associations represented as layered networks. Also in the 1970s the new field was formalized with the creation of the Cognitive Science Society and with the appearance of the *Journal of Cognitive Science*. The discipline focuses attention on information processing, as well as on the faculties of perception, memory, reasoning, emotions and language, investigating basic learning, and decision-making to mechanisms of high level logic and planning. The methods employed are those of the physical sciences, in ways that are most appropriate to the area studied, such as behavioral observations and experiments in the laboratory, mathematical and computer models of information processing and transformation, and brain imagery and other devices and methods borrowed from neuroscience, neuropsychology, and neurolinguistics. This represents a convergence from different directions on a single domain within the paradigm of a new science of the mind.

Anthropology also plays a role in this enterprise, exploring different cognitive maps in different cultures, the relationship between language, culture, and cognition, as well as in attempts to develop general theories of culture formation in co-operation with social psychologists, political scientists, and evolutionary biologists.

Every science passes through stages of development as it matures. First comes conjecture based on prior experience and sometimes innovative inspiration, leading to the formulation and testing of hypotheses and the systematization of research plans, operating at a mature stage within an accepted paradigm, a stage which Thomas Kuhn has called "normal science" and which he describes as essentially one of puzzle solving within the parameters of the paradigm with its own frontiers and with further conjectures and either eventual confirmation or refutation, and perhaps convergence on some other branch of science. Some critics imply that cognitive science is still in the conjectural stage. They admit that the computer model of information processing has been successful in explaining some aspects of learning, problem solving, and language use, but that too little attention has been given to the role of consciousness and the role of emotion in thinking, something that does not enter the picture in comparisons with machine simulation. They argue that the role of the physical environment in perception and thinking has been ignored.

Critics also say that the mind is a dynamic, not a computational system; that thought is inherently social, another factor which the current paradigm ignores, and that mathematical results show that human thinking cannot be computational in the standard sense, so the brain must operate differently, perhaps as a quantum computer which employs quantum mechanical effects in place of binary encoding, as is the case of digital computers which were the first inspiration for cognitive science. So a new convergence may be in the offing. It is not improbable to think about an eventual convergence of cognitive science, biology, and quantum physics sometime in the future, but for now that is still conjecture. Science advances from conjecture and refutation or confirmation, as well as with criticism and response. Response to criticism of cognitive science at this stage is that emotions, actions, and consciousness can be explained in terms of neural mechanisms, a field which will be further explored. Also, it is hoped that the development of computational models of social interactions will deal with the realm of cognition and social interactions.

Once the processes of thinking came back into the picture, the resources of different disciplines were brought to bear on a common point of interest, showing how convergence works and why a holistic approach reflects the overall goal of science; to understand and explain as much of nature as possible with the reasoning and the methods of inquiry characteristic of modern scientific thought and action.

THE DISCIPLINED STUDY OF LANGUAGE

Language is a species-specific property of human beings. It is cognitively based since cognitive abilities underlie language acquisition and use, and it is socially situated since it is acquired and used in social settings. Because cognition is seated in the brain, language also rests on a biological foundation. Language is also a social reality, not only because it is the principal means of human communication, but also because the acquisition of a language depends on social interaction, and because language plays a role in the formation and the maintenance of group and self-identity. Speech also serves to foster group cohesion. Moreover, language plays a major role in the shared perceptions of a community, perceptions that shape in some ways a community's particular view of the world.

Anything so central and so multifaceted can be, and has been, approached from different directions, with different theoretical perspectives, and employing different models, different methods, and varying research strategies, approaches which often proceed along parallel but sometimes overlapping lines. The same convergence trends seen elsewhere are also seen here, but without creating a new branch of science. Since this area of convergence, like cognitive science, involves anthropology, and since language is the topic of one of the four fields, we will examine it here in some detail.

The first disciplined study of language was in the form of *rhetoric* (the use of language for persuasive purposes), *poetic language* (its use for artistic expression), and eventually *grammar* (the attempt to explicate the inner structure of speech). Those disciplines were originally studied in order to use language more effectively, although they eventually became topics of philosophical inquiry. In the late eighteenth century attention shifted to the comparison of the grammatical systems in what were increasingly seen as related languages, for purpose of understanding their relationship with one another and their histories, thus creating the field of comparative and historical linguistics (see chapter 1).

The classification of the languages of the world also began at that time. One was a classification of different but related languages based on a common now extinct ancestral language or proto-language such as the Romance family of languages (Italian, Spanish, Portuguese, and Romanian) which are derived from Latin, and a typology based on structural similarities regardless of genetic relationship. In 1818 August Wilhelm Schlegel proposed a typology which is still used today in which he divides the world's languages into three types based on the number of morphemes (meaningful elements) which can appear in a single word and the modifications they might undergo. They are the analytic or isolating, the agglutinative, and the inflectional languages (Arlotto, 1972: 29–44).

Data in this enterprise came from written sources and that field of study became known as philology, the study of the nature of language and its role in culture as seen through the systematic study of written texts. In the nineteenth century an attempt was made to better understand language change by going out in places where language is used and directly observing linguistic variation in the form of the dialects of the spoken language. This developed into another field of language study known as dialectology which revealed even more about the broader reality of language.

In the late nineteenth century, Ferdinand de Saussure (chapter 2), a scholar of comparative historical linguistics, both narrowed and deepened the field by arguing that the task of a linguist is to study language structure which he called *langue* leaving to others the study of speech, *parole* (what people are actually doing when they use their language). He further made a distinction between synchronic studies of the language system (*langue*) at a single point in time, and diachronic studies which looks at how the structure changes over time. From those ideas there developed structural linguistics whose focus is synchronic, and the continuation of historical linguists, which is diachronic, a divergence that brought greater and more refined knowledge to both aspects of language. Noam Chomsky in the 1950s retained Saussure's basic dichotomy, calling *langue* linguistic competence and *parole* speech, and remaining focused on the synchronic nature of structure. His important innovation was a new formal model for describing linguistic structure known as generative and transformational grammar.

Dialectology continued alongside structural linguistics, focusing not on structure, but rather on points of variation in dialects, tracing their distribution over space and how they came about in terms of traffic patterns and geographic features in the linguistic landscape. The two branches of language study diverged to the point where in the 1960s Uriel Weinreich asked if the two could be reconciled in a paper titled "Is a Structural Dialectology Possible?" He thought they could and showed why.

Attention to the borrowing of words among languages was important in comparative linguistics, since the linguist had to know if the vocabularies of two or more languages under investigation had been retained from an ancestral form from which they both were descended, or whether those words had been borrowed from another source. Borrowing also became important in reconstructing the histories of individual languages as in the case of French in English, Arabic in Spanish, and so forth.

Some geographic areas are characterized by the use of different languages in the same wider area of interaction where lexical, phonological, and grammatical features diffuse across linguistic boundaries, even boundaries of different language families, forming what is known as a *Sprachbund* or convergence zone. The Balkans is an example of one such zone. Others are found in India and in Native North America and elsewhere. The study of such areas is known as aerial linguistics, a case of the creation of a new branch of language study by the overlap of dialectology and synchronic and diachronic structural linguistics.

In the late nineteenth century Romance philologist Hugo Schuchardt (1859–1864), working within the linguistic paradigm of his day, took an interest in a similar phenomenon, namely, language mixing. This line of research was resumed and intensified in the twentieth century in the study of pidgin and Creole languages, research which developed into yet another branch of linguistics which builds on the structural and historical lines of linguistic inquiry as well as attention to the social context of language use.

In 1964, Uriel Weinreich looked specifically at the phenomenon of language contact and founded the field of contact linguistics in which linguistic, social, and psychological factors are taken into account in borrowing. Five years later Einar Haugen, drawing on earlier studies of language borrowing, together with field work among the bilingual descendents of Norwegian immigrants in America, established the study of bilingualism which he and others have since developed.

Dialectologists had observed that not only geographic features play a role in the variation of single languages, but that social class and occupation were also important, thereby expanding the scope of dialectology. Thus diverging fields of language study—the social dimension of dialectology, areal linguistics, and studies of language mixing and pidgin and Creole studies—brought about a convergence on the social context of language which had been left out by structural linguistics.

A little before that, in the late 1950s, attention was increasingly turned to the relation of language and society, resulting in a convergence of linguistics and sociology and the emergence of a new branch of language study called the sociology of language and sociolinguistics. Those fields, usually called sociolinguistics, include the study of the correlation of the different speech varieties, or social dialects, with different socioeconomic classes within a language community. The work of Joshua Fishman and William Labov set this branch of research in motion. In 1968 Weinreich and Labov, together with dialectologist George Herzog, applied the perspective of sociolinguistics to the question of language change. They argued that the locus of change is in the interaction of people within ongoing communities with their built-in linguistic heterogeneity, and that the process of change can be seen in real time, and can be studied employing sociolinguistic methods and field observation. Thus different approaches in different fields converged to give a broader understanding of language.

The relationship of language and social context was also a major theme in the London school of linguistics, founded by J. R. Firth. This approach focused on the functional aspect of language following the lead of Bronislaw Malinowski who looked at language not in structural terms, but rather as a mode of social action. M. A. K. Halliday carried this work forward highlighting among other things the role of intralingual styles, or registers, which mark the formality or informality of the speech situation. Another topic in sociolinguistics is code-switching; switching from one dialect or language to another under different circumstances and for different purposes. In sociolinguistics the investigator identifies the varieties which comprise the linguistic repertoire of speakers—dialects, registers even different languages—and how they are used in varying situations.

Some sociologists approached language in a different way. They were interested in how language was used in social interaction and in carrying on routine negotiations among people. Harold Garfinkel pioneered this approach with what is called ethnomethodology, the direct observation of how people use language in everyday interactions. This meant that investigators were not limited to the sentence level, which forms the upper boundary of investigation for the linguist, but must take into account larger stretches of speech, or discourse. This approach, along with an interest in sociolinguistics, led to a new branch of linguistic study known as conversation analysis and discourse linguistics (also called text linguistics).

In the 1950s and 1960s, philosophers such as Ludwig Wittgenstein, John Austin, John Seale, and Paul Grice turned their attention to language, not in terms of its inner structure or its history, or its variation, or its place within a wider social context, but rather to what people do with language as a part of ordinary social interaction. Wittgenstein compared language use to the way people play games, and John Austin in 1950 wrote a book simply titled *How to Do Things with Words* where he viewed language in terms not just of sentences, but rather in terms of acts which realize the speaker's intention. Intentions and speech acts are a part of ordinary social interaction, such as greeting, commanding, inviting, asking, requesting, warning, threatening, and promising. Another philosopher, John Searle, worked out those ideas into a theory of speech acts in which he described such performative utterances as well as the "felicity conditions" which make them valid, for example conditions of intention and ability in which the speech act of promising creates a commitment and not a prevarication, a joke, or hyperbole.

Philosopher Paul Grice looked at another aspect of how people use language in their daily lives, an action that lies at the heart of coordinated linguistic interaction. He calls this the co-operative principle defined as the expectation that participants in a linguistic exchange each will make the required contribution to the interaction at the stage in which it is appropriate for expected purpose. There are no rules in this process, as there are in the production of speech sounds, words, phrases, and sentences but rather tacit agreement among speakers on a set of principles which people generally obey in the course of their everyday conversations.

These principles can be described in four maxims. The first is the maxim of quantity which says that you make your contribution to the conversation with no more information than is required for the situation at hand. The second, the maxim of quality, is do not say what you believe is false, or for which you lack evidence. The maxim of relevance is, keep your remarks relevant to the topic at hand, and the final maxim is that of manner, which means avoid obscure expressions and ambiguity, be direct (avoid unnecessary proclivity), and be orderly in your speech, otherwise you come across as unclear and bumbling. There are, however, occasions when it is best to flaunt those maxims, which we do all the time. For example, if a wife shows her husband a new dress and asks, "how do you like it?" it's sometimes best to flaunt the quality maxim and say, "fine" even if he might not think well of it.

Here we see how philosophy, which approaches language from a different disciplinary perspective with different questions, has complimented what structural and historical linguists and dialectologists and sociolinguists have said about that single strip or reality from their different disciplinary perspectives, each with its own questions, theories, and methods.

Since the 1960s the data storage and sorting capabilities of computers have greatly increased and have become more sophisticated, making it possible to construct databases, called corpora, which consist of large samples of millions of words taken from the real world of speech and writing in any domain of use, from which patterns can be identified. This approach, known as corpus linguistics, allows us to better understand the rules and conventions that underlie real life language.

This general widening of the scope of language study linguistic brings us closer to the ancient arts of logic and rhetoric and to traditional philology, but within a more unifying, cross-disciplinary perspective on language as a whole. This approach, which takes in discourse, conversational maxims, conversation routines, presuppositions, speech acts, speakers' intent, and context of situation, is called pragmatics. Here we see a case of convergence which now forms not only different branches of language study, proceeding from different fields—linguistics, philosophy, and sociology—but which forms part of the standard scholarly view of language.

One kind of discourse is narrative, a basic part of conversation along with dialogue, which is a primary way of constructing meaning and making sense out of something and communicating it to others. Labov addressed the question of form in this important stretch of discourse in the context of sociolinguistics, by examining narratives of personal experience in everyday conversation, showing that certain requirements have to be met in order for a narrative to successfully convey the information it is intended to transmit. He calls this aspect of structure in language use narrative syntax, which provides a model for the basic nature of narrative from which the other genres and styles have developed.

Another linguistically based analysis of discourse, both spoken and written, is stylistics, the study of the devices (rhetorical, grammatical, and phonological patterns) to achieve specific effects. Here we see a continuum from registers through oral storytelling to literary texts, like that from conversational narrative to different practical genres to high literature. Stylistics is based on the assumption that the primary procedures of textual interpretation are linguistic procedures, and that for authors to achieve the specific effect they have in mind they unconsciously, or consciously manipulate those linguistic devices. In studying style, as in analyzing narrative structure and function, scholars must check the relationship of one set of patterns to another across the text, with evidence for their conclusion, which then can be attested by analysts working with the same methods.

A more general view of function was formulated by psychologist and linguist Karl Bühler (18979–1963), known for his work in Gestalt psychology, who addressed the topic in a 1934 publication in which he identified three functions of language in terms of the typical act of speaking (*Sprechakt*). Which involves the speaker, addressee, and the external factors to which reference may be made by the utterance. This typology has figured prominently in many subsequent influential treatments of language. This tripartite division is: representation (*Darstellung*), also known as the referential, the denotative and the cognitive functions; expression (*Ausdruck*) also known as the emotive or effective function; and the vocative function (*Appell*).

In 1960 Roman Jacobson included Bühler's functions into a broader typology adding three more: the phatic, a term borrowed from Malinowski, the metalingual, and the poetic functions. The phatic function is the use of language for the sake of interaction. Examples are greetings and casual chatting about the weather or some other topic just to keep the interaction going, or chatter to sooth feelings or fear. The metalingual function is the use of language to talk about itself. This ranges from discussions in one to everyday conversation like "I don't follow you." "What do you mean?" or "Do you know what I mean?" and "What did you mean by that?"

Jacobson's sixth function is the poetic function of language which "focuses on the message for its own sake." This is the dominant, yet not the sole function of literature, and is used in activities other than verbal art (both written and oral). One example is for ease of memory as in such mnemonics as "thirty days hath September, April, June, and November, all the rest have thirty one . . . ," and in commercial and political advertising in the form of jingles and slogans such as the Eisenhower presidential punchy campaign slogan, "I like Ike," a three syllable phrase with equal stress on each syllable and with the repetition of the same vowel in each word. Such juggling of ordinary language makes this use of language stand out from the background of ordinary speech, thereby fortifying or amplifying the message.

At this juncture we see an overlap with beneficial mutual influences, between linguistics—the scientific study of the highly structured linguistic code—and the humanities, an overlap which was originated by the Russian Formalists who are associated with scholars like Victor Shlovsky, Valdimir Propp, Boris Eichenbaum, and Roman Jacobson among others, a school of thought which influenced such scholars as Baktin and the French structuralists working in linguistics, anthropology, and literary studies. This school emphasized the autonomous poetic nature of language and defined properties specific to it. It also advocated a scientific approach to literary studies instead of the less rigorous and more diffuse nature of literary criticism as it was then practiced.

Russian Formalism shifted its center to Prague in the 1930s under Stalinist repression. The Prague School of linguistics, whose work has influenced linguistics in general, was interested in both structure and function. They also studied the written language and described it as a system different from that of spoken language. Later the study of those differences would expand to not only the relationship between writing and speech, and how writing affects language in general, but also to the impact of writing on culture and thought with its distinction between orality and literacy (see chapter 11). The Prague School was also the first to examine language planning one aspect of the field of applied linguistics.

Function has become central to one formulation known as functional theory that holds that since language is fundamentally a tool it is reasonable to assume that it is best understood in terms of how it functions. Thus rather than defining grammar in formal terms, as do the structural and the transformational linguists, it is more appropriate to define the function performed by language and relating those functions to the linguistic elements that carry them out, such as the way syntactic functions as subject, object, and so forth are expressed, and how such elements as theme, topic, and focus are codified and employed within the language system.

Two other disciplines which have converged on a common topic is linguistics and psychology to form the discipline of psycholinguistics. This field involves the way a first language is acquired by children as they mature, and how language is produced and comprehended. In the 1950s Chomsky introduced a cognitive perspective on language when he defined it as "linguistic competence;" what a speaker must *know* about the linguistic structure in order to generate only the syntactically and semantically well-formed sentences of the language. He elaborated this line of thought in his book *Language and Mind*. Since then another view of linguistics has emerged called cognitive linguistics that argues that language is best understood as a part of cognitive processes, and that there is no autonomous cognitive system in its acquisition, production, and comprehension. Here we see a convergence of linguistics and cognitive psychology which are also constituent parts of cognitive science.

It is clear from this discussion of function that we are also talking about meaning, for example the referential function refers to denotative meaning, and the expressive function to affective or emotive meaning. This brings us to the field of semantics which for a very long time has been approached in different ways by scholars in different disciplines. Philosophers speculated on its nature and linguists looked at how words and stock phrases encode meaning in the form of prepackaged concepts, and how those units are combined by syntax into phrases and sentences. There is, however, still no overall theory of semantics which encompasses those different approaches into a single unitary theory.

Meaning, intention, and inference are all involved in another field of study known as semiotics in which the study of language fits as one part of the broader whole. Philosopher Charles Saunders Pierce used the term "semiotics" to describe that area of study, defined as the theory of signs and semiosis, any activity involving signs and the transmission through them of information. Pierce identified three kinds of signs: index sign whose meaning is a part of the thing it stands for, icons that derive their meaning by virtue of their similarity with what they stand for, and symbols which derive their meaning from convention, the fact that everyone agrees that's what the sign means.

Semiotics also consists of three kinds of analysis: syntax, the relationship of signs with one another within the system; semantics, the relation of signs with what they signify outside the system; and pragmatics, how signs are used. Language is primarily a symbolic system yet index signs are a part of speech as when differences of intonation indicate the emotional state of the speaker, and icons appear in the form of onomatopoeia as in words such as the "hum" of a bee, the "tweeting" of a bird, and the "pop" of a punctured balloon.

Pierce described this process from a philosophical perspective. Saussure described the same thing, which he called semiology, from the perspective of social psychology and thus more broadly psychology, and Thomas Sebeok brought them together as aspects of the same field of study. Semiotics is not only applicable to conscious communication in human but also in animal behavior, the latter study in zoosemiotics, and on the cellular and molecular levels as biosemiotics. In the 1940s a new, wholly independent way of dealing with signals and information transmission was developed by mathematician and cryptographer Claude Shannon and Warren Weaver with help from Alan Turing, an influential figure in the development of the computer.

Shannon and Weaver devised a model to account for communication which consists of the information source; the transmitter, which encodes information into signals; a channel, in which the signal is transmitted; a receiver that decodes the message in the signal; and a destination where the message arrives. An adaptation of that model for linguistic communication involves the speaker (or writer) who produces the encoded signal and the hearer (reader) who receives it. Both must have the intention to communicate. They must also share the same code (language) in which the message is encoded and sent as a signal through some channel (auditory in the case of speech and visual in the form of writing and sign language). The hearer then infers the speaker's communicative intention on the basis of his knowledge of their linguistic code, within the context in which the transaction takes place (the context of situation) and against a background of shared knowledge, thus organizing all of the processes of linguistic communication in a single model which shows the relationship between the different approaches to the study of language as summarized above.

The ability to acquire and use language is cognitively based, and cognition in turn is rooted in human biology. Those faculties, therefore, are as much a part of the genetically determined nature of humans as the abilities to breath, then walk on two legs and to see and hear. Studies of language acquisition across languages and cultures have revealed a consistent pattern in the developmental stages of first language acquisition which correlate with motor milestones such as crawling, then walking and running as the brain develops.

The field of neurolinguistics has emerged to study the neural mechanisms that control the acquisition, the production, and the comprehension of language since the functions of language are primarily an aspect of electro-chemical activity in the brain. Non-invasive imagery has provided evidence of brain activity in speech. Blood flows are also cues as to neural activity; by imaging, patterns are recorded of magnetic field variation due to that flow. Devices used in this line of research—MRI (Magnetic Resonance Imagining), PET scans (Positron Emission Tomography), and EEG (electrocephalograms)—provide scientists with a clearer picture of the neurological basis of language.

In sum, language is based on cognitive abilities, which in turn are grounded in the neurological structure of the brain. Language is also integral to social interaction and to culture, serving as a medium of everyday communication, of contemplation, and of artistic expression. It has therefore attracted the attention of investigators across the disciplines—from the physical and the social sciences to the humanities—each asking its own questions, each with its own research designs and its different theoretical directions, and each with its own bodies of accumulated data, all converging on the same realm of reality while the remaining separate disciplines give us an ever-broader understanding of what it is and how it works: an example of a holistic perspective where knowledge of the whole is increased through knowledge of its various parts.

SEVEN

American Anthropology and the Formation of the Scientific Holistic Approach

In the previous chapter, we examined the convergence of different disciplines on common areas of interest to show that the kind of holism which characterizes anthropology in its four field approach is by no means unusual but instead leads to a better understanding of the phenomena under study. In the chapter before that we discussed a meaning of science which accommodates such holistic views. We spent some time examining that broader meaning and its philosophical foundation, since American anthropology is in fact a part of that intellectual tradition, brought to an already thriving American anthropology by Franz Boas who shaped the discipline into what it is today. In that regard, historian of anthropology Adam Kuper has said, "later American students did not always appreciate how deep were the German roots of Boas' anthropology" (1988: 125; see also Stocking 1996). Nor do they realize that some of the leading and most influential anthropologists of the first generation (Robert Lowie, Edward Sapir, and Alfred Kroeber) were also involved with that stream of thought in the formative stage of the development of contemporary American anthropology.

Franz Boas (1858–1943) was born in Minden, Germany, in 1858 where he graduated from the Gymnasium. He then studied at the universities at Heidelberg, Bonn, and Kiel where he was given a thorough education of his time in mathematics, the sciences, languages, and the humanities. He was first drawn to the study of physics and mathematics, receiving his doctorate in physics in Kiel in 1881 with a dissertation titled *Beiträge zur Erkenntniss der Farbe des Wassers* ("the Recognition of Color in Water") which dealt with the adaption of light in water.

This beginning in physics contributed greatly in determining the direction of his intellectual career and which, as Alfred Kroeber observes, helped him achieve greatness in other fields as well, something which Kroeber says was rare in anthropology (Kroeber 1943: 7). Boas was also interested in the mind and in culture. His thesis on the perception of color led him to the study of psychophysics and from there to cultural geography and from there to anthropology, in rapid succession thus a straight line from the physical to the human sciences in a short time.

Boas was not interested in experimentation. This left two courses open to him: theoretical physics with maximum amount of mathematics but a minimum of phenomenal experience, or switch fields to tackle phenomena without experimentation but, if possible, with a mathematical approach. He chose the latter, not simply applying physics to human phenomena but rather transferring the essence of its method so far as it was applicable to his area of interest (Kroeber 1943: 7).

Boas's first field experience was in Baffinland between 1883 and 1884. Afterwards he went to Berlin, where he became an assistant in the Museum für Völkerkunde (Museum of Ethnology) founded by Adolph Bastien and at that time still under his direction. There Boas came into contact with other assistants and with others whom he met at the Gesellschaft für Ethnologie (The Society for Ethnology) which, says Kroeber "proved intensely stimulating" (Kroeber 1943: 9). It was in 1885–1886 that he probably began his association with Rudolph Virchow "who perhaps influenced him more than any other scientist" (Kroeber 1943: 10). In the summer of 1886 Boas "habilitated" at the University of Berlin and received there an appointment as Docent in Geography. During that time he was influenced by geographer Friedrich Ratzel, but he eventually left from that branch of geography since it went no further than environmental influences while Boas's interests were more far reaching.

We see here a man who in his early intellectual life was exposed to the rounded education of the time, with a center in the physical sciences and the opportunity to expand from there to other fields, and in the process to make contact with the best minds of the times in Germany in various fields. Moreover he maintained contact with developments in that tradition while he was in the United States, through the literature and through frequent trips to Europe. Robert Lowie would later say that there is no "easily condensed and soul-stirring scientific message" in Boas's work (Lowie 1937: 131), yet he did write one paper which gives a clear indication about his thinking concerning the scientific method and its application to both the physical world and to the human sphere within it.

That paper was written in 1887 for the journal *Science* (later reprinted in *Race, Language and Culture*) titled "The Study of Geography" which George Stocking says was justification to physical scientists for his systematic study of material that was not subjected to quantitative determination or laws. Boas argued in that paper that both the search for laws and an attempt at "understanding of human phenomena for their own sake" are legitimate when approaching the problem of scientific knowledge. Stocking says that it is not clear whether or not Boas at this point had read Dilthey, whose *Einleitung in die Geisteswissenschaften* appeared in 1887, but that Boas was later certainly familiar with Dilthey's work as seen in an article he published in the *American Anthropologists* in 1907 (Stocking 1965: 64). The closeness of what Boas was thinking and doing and what Dilthey and the neo-Kantians were writing indicates that Boas was fully within that intellectual stream and not just responding to the ideas of others as they came along, for as Matti Bunzi points out, the frequent parallelism of Boas's argument with those of Dilthey and the neo-Kantians, extended to similarities of phrasing and similar in some cases to sources not yet available to Boas (Bunzi 1996).

"The students who were attracted to this exotic discipline in the United States at the beginning of the twentieth century," says Kuper, "were overwhelmingly of German extraction. Most of them had come to America as children, but their homes and schools were in New York" (Kuper 1988: 140). Among them was Robert Lowie (1882–1957). Lowie was born in Vienna and came to the United States when he was ten years old. His family settled in a German section of New York City where he grew up. German was the language of the home, and his culture was that of German-Americans. He inherited his maternal grandfather's library which included books by German philosophers. Like other prominent anthropologists of his time his scholarship originated in the German idealist and Romantic tradition of Kant, Hegel, and Herder. In his history of anthropology of thought (1937), Lowie thoroughly explored the role of German ethnology in the development of the discipline, especially the contributions of Waitz and Klemm.

Edward Sapir (1884–1939) was not only a prominent and influential anthropologist, but one of the foremost linguists in the country who helped establish linguistics as an independent discipline. Sapir was born in German Pomeranian and spoke both German and Yiddish as a child. His parents immigrated to the United States when he was five years old and settled in New York City. He studied German linguistics at Columbia where he fell under the influence of Boas and switched to the study of American Indian languages. He, along with Boas, developed the relationship between anthropology and linguistics and he was the leading figure in linguistic anthropology at its inception. Not only was Sapir versed in the German linguistic tradition, but he also kept up with the intellectual currents of the time. Lowie says that Sapir paid Rickert a compliment when he wrote, "for a penetrating analysis between historical and natural science . . . .I strongly urge all anthropologists, and social scientists generally, who are interested in method to refer to Heinrich Rickert's difficult to masterful book *Die Grenze der Naturwissenschaftlichenbildungsbegriff*. I have been greatly indebted to it" (quoted in Lowie 1956: 1007).

Another anthropologist of that generation was Alfred Kroeber who has been described as the dean of American anthropology, "probably the greatest general anthropologist." Kroeber was a prolific field worker and a theoretician who worked in all fields of anthropology, except physical anthropology. He was born in New Jersey but the family moved to New York when he was young. His father had emigrated from Germany with his family when he was ten years old and his mother was of German extraction. The family was bilingual and German was the language of the home. Kroeber had a private tutor who has been described as a vigorous German teacher, and his formal schooling was in the German pattern (Steward 1973).

His reading and thinking on the subjects of culture and history ranged widely, and included a sound knowledge of the German intellectual tradition, especially the work of the neo-Kantians and more specifically that of Heinrich Rickert. Kroeber approved of Rickert's redefinition of the *Geisteswissenschaften* to the *Kulturwissenschaften* but he disagreed with the dichotomy implied in Rickert's work. Kroeber tells us that he had based his own view of the relationship of culture and science on Rickert's work, but that he has gone beyond it by arguing for the unity of science rather than sharp dichotomies. In this respect Kroeber also acknowledges the work of Max Weber.

Another major contributor to anthropology was Bronislaw Malinowski (1884–1942) who has been described as "a major contributor to the transformation of nineteenth century speculative anthropology into a modern science of man" (Metraux 1968: 541). Like Boas, Malinowski started out as a physicist, taking his PhD in physics and mathematics from the University Krakow in 1908, before turning to the study of social and cultural variation. He viewed anthropology as a field-oriented science "in which theory and the search for general laws must be based on intensive empirical research involving systematic research and detailed analysis of actual behavior in living ongoing societies" (Metraux 1968: 541). He viewed culture as a universal phenomenon, but whose many variations should be studied in all their individual peculiarities in a systematic manner, which in turn would open the way for systematic cross-cultural comparisons. The introduction to his ethnography *Argonauts of the Western Pacific* might be regarded as the first field ethnographer's handbook.

His view of anthropology as scientific was explicitly spelled out in a collection of essays that appeared in 1944 titled *A Scientific Theory of Culture and Other Essays*. In the preface, Huntington Cairns notes that Malinowski's training in mathematics and the physical sciences gave him a firm grasp of the basic elements of the scientific method. At the same time the breadth of his interest remained free of the dogma usually associated with the physical sciences. "His own practice," writes Cairn, was "in strict accord with the contemporary standards which require a meticulous knowledge of the whole life of individual tribes" (Malinowski 1944: vi).

Malinowski was from Poland, yet from that part of the country which at the time was under the rule of the Austro-Hungarian Empire. His intellectual grounding was therefore within the broader sphere of German intellectual influence, an influence which is clearly seen in his work and in his comments on science. After studying physics, he studied economics at the University of Leipzig under Karl Bücher, one of the founders of non-market economics, and psychology under Wilhelm Wundt. Although Malinowski was not concerned with archaeology and physical anthropology, his theoretical writings clearly reveal a view of science which bridged the gap between social phenomena and the kind of cultural phenomena usually studied in the social sciences and the humanities.

Thus we see that the more inclusive idea of science as worked out on a sound philosophical base in the German intellectual tradition was well established in the minds of the major figures of anthropology at the beginning of the twentieth century, men who developed the discipline as a holistic science in both theory and practice.

EIGHT

Physical Anthropology

Physical, or biological, anthropology is the study of humans as one species among others in nature, thus representing the physical science end of the continuum which spans the holistic discipline of anthropology. This subdiscipline includes the study of the genetic composition of the species, its evolutionary history, the biological features of living human populations, as well as the study of intraspecies variation and the relationship of humans and other primates.

The origin of physical anthropology goes back to the late eighteenth century with the work of Johan Blumenbach (chapter 1). The field was advanced in the latter half of the nineteenth century by two pioneers in human biology, Paul Broca (1824–1880) in France, and Rudolph Virchow (1821–1902) in Germany. Broca, founder of the Anthropological Society of Paris (1859), is best known for his study of the brain. In this context he was also interested in the skull, and from that interest he developed the field of craniometry, the precise measurement of skulls, a method used in the early days of anthropology for the study of human physical variation.

Virchow on the other hand emphasized research on environment and disease in the human body. He was also interested in the general biological aspect of humans, as well as their culture and their culture history. To promote those lines of research he founded, in 1869, a society for physical anthropology, ethnology, and prehistory which foreshadowed the four fields approach which emerged in the United States. Today in Europe those lines of research are carried on separately. The fact that they converged in this country in the way they did is thanks to the vision and the efforts of Franz Boas, who came from the tradition in which Virchow was a major figure, and as well as to the work of Ales Hrdlicka (1869–1943). Hrdlicka was a physician who had studied anthropology in France, and who in the United States founded the *American Journal of Physical Anthro-*

pology. He was also first curator of physical anthropology at the Smithsonian Institution (1903).

In the late nineteenth and early twentieth centuries, interest in American physical anthropology was focused on racial classification and on the discovery and the classification of fossils in the chain of human evolution. By mid-century, especially with the work of Sherwood Washburn, interest shifted from classification to the study of evolutionary processes. With advances in genetic research a new branch was developed, anthropological genetics, which attempts to understand the genetic makeup of human biological variation and which, together with paleoanthropology, strives to better understand human evolution. One aspect of this area is molecular anthropology which employs DNA analysis to determine relationships between human groups as well as evolutionary links between ancient and modern populations. Another branch of physical anthropology is paleopathology which studies disease, trauma, nutritional disorders, and occupationally derived stress as seen in bones and mummified tissue and in the stature and morphology of human remains. And bioarchaeology is the study of past human cultures through the human remains recovered in archaeological sites.

We also see in the study of human evolution the co-operative effort of different branches of science such as paleoclimatology, primatology, ostology, ethology, evolutionary biology, evolutionary psychology, linguistics, embryology, and genetics. Another discipline that overlaps physical anthropology is human behavioral ecology, the study of behavioral adaptations to the environment, reproduction, and human development and genetic responses to environmental stress. Primatologists study the morphology, genetics, and behavior of non-human primates as a way of inferring how and why similarities of human traits have evolved. We thus see a natural convergence within anthropology and with neighboring disciplines in the study of the biological nature of humans.

PALEOANTHROPOLOGY

Central to the search for evolutionary roots and the course of human evolution is paleoanthropology, the study of fossil evidence of extinct hominids and other primate species. For this task, anthropologists deal with geological time, in this case the Cenozoic Era, the time when the first primates appeared. This era extends back from 58 to 65 million years, and is divided into epochs.

The taproots of the human family tree lie in Africa in the Middle and Late Miocene when Australopithecus diverged from the ancestors of chimpanzees, bonobos, and gorillas. The earliest species known in that line, dated at 4.5 to 5.5 million years ago, is *Australopithecus ramidus* whose fossil remains were discovered at Aramis, Middle Awash in Ethi-

Table 8.1. Chronology

The Cenozoic Era		
Epochs	*Dates (in millions of years)*	*Primates*
Holocene	0.01–Present	Modern human
Pleistocene	1.8–0.01	Development of Humans
Pliocene	5–1.8	Austrolopithecines and *Homo habilis*
Miocene	25–5	Variety of Apes and very early humans
Oligocene	35–25	First monkeys and apes
Eocene	58–35	Prosimians similar to lemurs and tarsiers
Paleocene	65–58	Earliest primates

opia. *A. ramidus* was bipedal but with a grasping foot which indicates he was arboreal. Evidence dating from around 2.5 million years ago from Gona, Awash, Ethhiopia, discovered by Sileshi Semaw has linked stone tools to what he calls *Australopithecus garhi*. And recent finds from a site near Lake Turkana in northern Kenya indicate that stone tool-making may go back some 3 plus million years, although this is still hypothetical.

The first known member of the genus Homo is *Homo habilis* which was descended from Australopithecus, but with a brain a good deal larger, somewhere around 650 cc compared with the 450 cc size of Australopithecus. *Homo habilis* coexisted with a contemporary form of Australopithecine known as *Australopithecus boisei* (or *Paranthropus boisei*) which eventually disappeared. *Homo habilis* also used stone tools, an innovation along with its less specialized diet that made it possible for the species to survive and to become the precursor of a new line leading to modern humans.

The next million years was a time of rapid increase in encephalization (the brain mass related to the total body mass of the animal) which led to the next stage in the evolutionary process with the appearance of *Homo erectus* whose brain had doubled to 850 cc. *Homo erectus* used more sophisticated tools, and is believed to be the first to use fire, an innovation which appeared about a million years ago and is unique in the animal kingdom. This was the species that left Africa and ranged across Eurasia, its remains having been found in Java and Peking.

Homo erectus was also the ancestor of *Homo heidelbergensis* (also sometimes called *H. rhoderensis*) which originated in Africa 800,000 or 1,300,000 years ago and lived until 200,000 years ago in Africa, Europe, and Western Asia. The African branch of *Homo heidelbergensis* is thought to be the ancestor of *Homo sapiens* which appeared 350,000 and 400,000 years ago. From the Eurasian branch evolved Neanderthal some 250,000 and 300,000 years ago. Another species of the genus Homo called the Denisovans has been discovered in Western Siberia, a relative of Neanderthal which branched off from the human lineage 600,000 years ago and from Neanderthal, probably in the Middle East, 200,000 years later. Modern humans may have begun to migrate out of Africa between 90,000 and 130,000 years ago. They possibly entered Europe around 43,000 to 45,000 years ago, where they encountered the Neanderthals, and in Asia, the Denisovans. A detailed study of the genome reveals a complex pattern of interbreeding between the three species, which in the case of *Homo sapiens* probably occurred in the early stage of their migration from Africa.

The record of human evolution, briefly summarized above, has been pieced together from fragmentary fossil evidence aided by stratigraphy and laboratory based dating methods and with the help of genetic analysis with certainly more to come. The next step in understanding the course of evolution is to establish, on the basis of geological evidence and laboratory tests, under what conditions those evolutionary transformations took place. An important factor here is climate fluctuations, a natural part of planetary dynamics, and the environmental factor that determined the nature of the habitat and resources which might account for the evolutionary process as indicated in the fossil record.

THE EVOLUTIONARY PROCESS

The period during which the genus Homo separated from Australopithecines was a time of climatic variability characterized by a long term drying trend interspersed with shorter intervals of moister. This condition was discovered by drilling off the African coast to a depth of one hundred meters. In that way columns of layered silt are revealed in which lime is mixed with dust from the continent that has blown over the sea and that has settled on the ocean floor, acting as time capsules, for in each layer are found isotopes, trace elements, and suspended solids from plant life which make it possible to reconstruct ancient African climatic changes going back seven million years.

Such analyses revealed that during the period of separation what once had been dense forests had thinned out into lightly wooded areas alternating with open grasslands, changes for which each species had to develop its own survival strategy. The abrupt and deeply felt climate fluctuation may have also stressed the survival strategies which had been worked out by different species. Under such conditions some species became extinct while others responded with innovations which led to evolutionary changes. This is what may have happened in the case of the genus Homo.

One theory holds that with the change in the environment the predecessors of humans lost their ability to climb, and perfected bipedal gait, so that three million years ago Australopithecines resembled an upright chimpanzee that walked on two legs. The remains of *Australopithecus ramidus* however indicate that the species Australopithicus was already bipedal. At any event they lived now not in lightly wooded country sleeping in trees but rather on the savanna where they had to eat savanna plants as well as roots and tubers, rather than their customary diet which until then had been like that of modern great apes. In the course of time they added meat to their diet. The diet can be ascertained by an analysis of isotopes found in the fossil which makes it possible to show whether the individual consumed vegetables or if they also ate meat.

Teeth adapted to a vegetable diet were inadequate for meat eating, and thus objects outside the body were used to tear the flesh for consumption, and so the further development of rudimentary tools. It took another 700,000 years, however, for the first hand axes to appear as seen in finds in the Olduvian culture. Better tools gave them access to a more efficient diet which in turn nourished the evolving brain. In this way the ancestors of humans had come upon a way of departing from the biological nature of the great apes.

It was in that new environment that *Homo erectus* appeared. The most marked evolutionary changes seen at this stage are an increase in the size of the brain and the shape of the teeth that indicate that they now clearly ate meat. This change in diet affected metabolism which meant that the digestive tract shrunk and the muscles lost their former strength which in turn advanced the development of the brain, both changes accounting for a large part of the energy consumption of the organism. Each calorie saved in that way was diverted to the brain which increased in size, and with this growth there was an increase in the intelligence of the species, an innovation which augmented its adaptive capability. About 1.9 million years ago the highly successful hunting and gathering strategy was developed whereby the plant and animal resources available in the environment were efficiently exploited, creating an ecological niche that would remain a characteristic of the genus until the Neolithic Revolution around 12,500 years ago when plants and animals were first domesticated, a strategy that persisted in places well into the twentieth century.

The swift and sudden increase in brain mass cannot be accounted for solely by the addition of meat to the diet and to the strikingly successful hunting and gathering strategy, but also to another highly significant achievement of *Homo erectus*, the discovery of how to control fire. Fire as a "tool" not only served as protection against predators, for *Homo erectus* lived not in trees as had its ancestors but rather on the ground where they were more vulnerable, but also because cooking and roasting increases the caloric intake of meat, and thus increases the energy derived from the diet which in turn contributes to the growth of the brain. The problem in this scenario, however, is the date when the control of fire began. One hypothesis is that it took place around two million years ago, yet there is no proof of that in the archaeological record. Anthropologists, however, are confident that such proof will eventually be discovered.

Neanderthals, like *Homo sapiens*, were hunters and lived in social units. They also buried their dead and used symbols and probably also used language. The question is, why did the Neanderthals disappear with the arrival of modern humans? One theory is that competition with humans under conditions of climatic stress led to their ultimate extinction. In that respect, Pat Shipman, citing biological research, says that when a new species enters a region the greatest competition among the native species is felt by the newcomer's closest relative, in this case Neanderthals. She takes this a step further with an interesting hypothesis in which the early co-operation of the two apex predators, humans and the dog, provided the competitive edge that led to the dominance of modern humans and the disappearance of the Neanderthals.

The full domestication of the dog came much later than the extinction of the Neanderthals, but the discovery of what Shipman calls "wolf dogs" shows that there was a species at the time which was not a dog and not a wolf, but which had the abilities of both in terms of speed and endurance in running, and a keen sense of smell. Bones of those animals have been found in association with what she calls "megasites," places where large deposits of bones from the woolly mammoth were found along with signs of human activity. This suggests to Shipman an association in hunting between the two predators.

The wolf dog would run a large animal to exhaustion or hold it at bay until the human hunters, unable to pursue in that manner, came up for the kill. The hunters would then quickly strip the meat that they needed, and the wolf dog would eat the rest. Neanderthals also hunted but, like humans, could not run down game as did the wolf, and were thus less efficient in the hunt than the man-dog combination. Moreover Neanderthals had to move in on the prey for the kill with stabbing weapons, which could be perilous. Human hunters, however, could stand off at a distance and kill their prey by throwing spears or with bows and arrows, a less dangerous way of bringing down game. The co-operation of those two predators, man and dog, thus gave *Homo sapiens* a survival advantage, resulting in the eventual extinction of the next closest species in the environment. If Shipman's hypothesis is correct then we see a biological first in which two otherwise competing species (early humans and wolf-like predators) co-operated for their mutual benefit. This would also provide an explanation of how the domestication of the dog eventually came about long before the domestication of other species, an event which played such an important role in human cultural development.

EVOLUTION AND LANGUAGE

A major distinguishing factor of humans is their use of language. The origin of language, however, is one of those murky areas in human evolution, for the only fossil remains that might indicate the presence of language is the size of the brain, the organ in which language is based, as seen in the cranial capacity of the fossil. Here is where anthropologists look to primatology for a possible way of solving the problem, for given the close genetic relationship between humans and the great apes, it was thought that something could be learned about the evolution of language by comparing their communication abilities with those of humans.

Speech is impossible with the great apes due to the nature of their pharynx; however, they have a communication system based on calls and gestures, some of which are like those used in human communication. Experiments were therefore conducted to see if chimpanzees could communicate using human language through sign language and other devices. The experiment to teach the chimpanzee Washoe American sign language is one well known example. Washoe was able to spontaneously combine hand signals to form new words based on the use of 130 signs. One of the most successful experiments of this kind was with the gorilla Koko, who combined words creatively to name new objects. Other experiments were carried out using lexigrams rather than signs, that is, symbols representing common objects, verbs, and moods.

Those systems are in part iconic, thus it is doubtful that apes have the capacity to handle a totally arbitrary symbolic system such as language. Moreover those experiments were conducted under specific conditions and required immense effort in contrast to the ease of language acquisition of children. The examples of apes that have learned to communicate in a human manner is analogous to behavior taught to performing animals by trainers, behavior which is not passed on to the next generation but is a product of the close interaction of the trainer and the performing animal under highly artificial conditions. Thus despite the achievement of some test animals in communicating by means of some aspects of human language, there is a wide gap between apes and humans in terms of communication and of the cognitive abilities that underlie it, products of the different courses of evolution followed by the ancestors of the higher primates and mankind.

Dereck Bickerton (1995) offers a scenario of how language may have emerged from some early primate communication system. He points to the relatively slow rate at which tool making changed over the first four million years of hominid evolution. He cites the small number of artifacts in the caves of Zhoukoudian in northern China which were continuously occupied from about 500,000 years to 200,000 years ago. The lack of change during that time reveals a long period of cultural stasis which cannot be associated with the possession of full language.

By circa 100,000 years ago, the time when Neanderthal Man lived in the Middle East and Europe, the cortex of both species had attained approximately its present size. It is therefore quite possible that they both had the mental capacity for complex language. Some cultural evidence, such as the deliberate burial of the dead and sophisticated tool making, also indicates that both species had developed modes of communication and abstract reasoning. It was thought at first that Neanderthals lacked a pharynx necessary for speech. Then in 1989 a Neanderthal skeleton was discovered in Israel with the small bones of the larynx still intact indicating the individual had the physiological capacity of speech.

With this line of evidence and reasoning, the theory as to the evolution from a primate call system to proto-language is that as social groups became cohesive in a human way around the need for subsistence and defense, communication became more important and thus set in motion a process in which language, culture, and the brain all evolved together, each stimulating and reinforcing the other.

SEARCH AND DISCOVERY

This picture of human evolution was pieced together from small and fragmentary, sometimes inadvertently discovered, fossils with gaps and open questions, creating a frontier of knowledge on which each new find can lead to new avenues of research and potentially major changes in evolutionary history; a process that holds a special fascination for the general public and which fuels the imagination and the motivation of paleoanthrologists. The major find which initiated this field of inquiry was the discovery of Neanderthal Man in 1856 (chapter 2). Another major find was that of *Australopithecus africanus*, which like Neanderthal Man was an inadvertent discovery. It was made in 1924 when blasting in a cave in South Africa by the Buxton Lime Works exposed some fossil remains. Robert Burns Young, who was employed by the company, sent the skull to Raymond Dart, professor of anatomy at Witwatersrand University in Johannesburg. The fossil was called the Taung Baby since it was found in the Taung District in the Southeastern margin of the Kalahari Desert. Since then three hundred pieces of the Australopithecus species have been found in East Africa.

Some finds are inadvertent and close to home such as the cases of Neanderthal Man and *Australopithecus africanus*. Others are the results of planned projects with the goal of discovering fossils in a likely place. Such was the case of the discovery in 1964 of *Homo habilis* by Louis Leaky and his family in the Olduvai Gorge in Tanzania, a part of the Great Rift Valley that runs through East Africa. Their meticulous work lasted several years and was highly significant for not only did it yield new specimens to fill in the gaps in evolutionary history, but it also radically altered ideas about the time scale in human evolution. The dating of fossils and tools found in that way involve radiometric methods of dating the layers in which fossils and artifacts are embedded, as well as the use of potassium argon and argon-argon dating.

Another find by an organized expedition was the well-publicized discovery by a team of anthropologists of *Australopithecus afarensis*. The fossils were found in 1974 near the village of Hadar in Ethiopia and were identified as the remains of a female that lived 3.5 million years ago which have been called *Australopithecus afarensis*. The size of the find was unusually large consisting of several hundred pieces of fossilized bones, in all 40 percent of the skeleton. It was especially important for the abundance of material which provided much information about that phase of evolution that was then unknown.

A part of the adventure of paleoanthrology is the discovery, which includes the romantic aspect of physical anthropology: finding something new in often distant, remote and inaccessible places, sometimes involving hardship, a find that may radically change an established theory and set research in a new direction, giving us a clearer picture of how we came about as a species and how we evolved over time. For example the excitement was high when the fossils of *Australopithecus afarensis* were found. One of its leading figures on the team was Donald Johanson, who tells us about the sense of discovery when the remains first came to light. He writes that "As a teenager, I dreamed of traveling to Africa and finding a 'missing link' but never in my wildest fantasies did I imagine that I would discover a fossil as earthshaking as" this one.

He describes the day of the discovery; waking up in his tent before dawn, looking at the stars glittering through the mosquito net, listening to the morning sounds of camp life, and later "the scaring noonday sun beating down on my shoulders" and the discovery "the shock of seeing small fragments of bone." He goes on to say that "Most dedicated fossil hunters spend the majority of their lives in the field without finding anything remarkable. And there I was . . . a thirty-one year old newly minted PhD, staring at my childhood dream at my feet." The team celebrated that evening while the camp tape recorder played over and over again the Beatles song "Lucy in the Sky with Diamonds": so the 3.2 million year old fossil they had just discovered was named Lucy, a name now known everywhere (Johanson and Wong 2010).

THE LATEST DISCOVERY: *HOMO NALEDI*, SOUTH AFRICA

In September 2013, two spelunkers (recreational cave explorers) Rick Hunter and Steve Tucker were exploring the inner region of the Rising Star Cave in South Africa when they found a narrow opening, so narrow that the spelunker had to extend one arm in front of him and the other close to the body to wriggle through; the superman crawl as they call it. He was only able to enter the opening because he was thin. The narrows descended into a larger chamber which, to his surprise, was littered with bones. The two spelunkers reported their find to Lee Berger at the University of Witwatersrand in Johannesburg. The first fossils of Australopithecus had been discovered in South Africa, but since then the search for the origin of *Homo sapiens* had shifted to East Africa. Berger, however, was convinced that South Africa had more to reveal in that regard, and so he was very interested in what Hunter and Tucker had found.

He sent out a message on Facebook calling for paleoanthropologists worldwide to assist in the recovery of the fossils. He received fifty-seven responses from which he chose six, mostly women slender enough to enter the tight space to the cavern. They retrieved 1,550 bones and teeth, more than all the finds together in South Africa, consisting of fifteen individuals, many lying in the same position where they had been left. Some had been disturbed by water and sediment entering the cavern, and there were a few bones of mice and birds that had somehow found their way into the area, but otherwise the fossils were undisturbed.

Berger set up a base camp outside the cave where a larger team began the process of analysis. They found that the skull shape of the individuals from the cave was similar to that of the genus Homo but significantly smaller than that of *Homo habilis,* more similar in that regard to the skulls of Australopithecines. The skeletal anatomy revealed that the body mass and stature of the individuals were similar to small-bodied humans. For those reasons Berger and his colleagues have classified this species as a hominin which they have named *Homo naledi*. Others, however, think it is better to wait for more evidence before classifying this newly discovered species.

So far the fossils have not been dated since they were laid out in what appears to have been a burial chamber with no contextual evidence such as strata in the soil by which they can be more easily dated. This presents a problem in the way evolutionary history so far has been written since burial, as far as the available record shows, did not occur until much later in the evolution of the genus. One line of thought is that rather than a single line from Australopithecines to *Homo habilis* and from there to *Homo sapiens*, that the course of the river of evolution meandered, with many branching streams that eventually became extinct along the way. This new find, therefore, may be one of those branches that evolved to some extent in a parallel fashion before disappearing.

What this find shows, however, is how spotty is the record which Nature has left us of our evolutionary past, and how new, sometimes unexpected finds, can still be made, opening new avenues of research leading to new theories of human evolution. It is there that paleontology offers the same kind of excitement in search and discovery of new evidence and new ways of thinking that was described by Donald Johansson with the discovery of Lucy, and which characterizes work in other branches of anthropology, in archaeology, ethnography, and linguists. This sense of excitement and adventure was expressed by one of the young members of Berger's team, Lucas Delezne, newly appointed professor at the University of Arkansas, when he said of his participation in the project that it was "a paleofantasy come true. In grad school you dream of a pile of fossils no one has seen before and you get to figure it out" (Shreeve 2015: 42).

NINE

Archaeology

Archaeology can be broadly defined as "The systematic study of antiquities as a means of reconstructing the past" (Fagan 1970: 3) which makes it, along with geology and paleontology, a historical science since they all deal with the temporal dimension. Archaeology can be pursued within different fields of scholarly inquiry. One such field is the classics, the study of Ancient Greek and Roman Civilization where archaeology is employed, together with the written record and its rich literature, to examine the sometimes spectacular ruins, sites, and objects left scattered across Europe, the Near East, and North Africa by the ancient Greeks and Romans. Archaeology can also be used in conjunction with the history of any society in any period, for filling in the gaps in the written record, and for clarifying, corroborating, and correcting it by reference to the material record of a time and place, and at times by revealing the bias of historians.

It can also reveal information not contained in the written record such as the material conditions of life which were not recorded, such as the manner of construction and the materials and dimension of houses as well as the recovery and examination of their contents. Material remains, systematically excavated, can also reveal patterns of agriculture, details of economic life, and how trades were carried on by the tools that workers left behind, facts which are ignored in the documents of the times.

Written history accounts for only a fraction of the human past, which means that archaeology is our only means of understanding what went before there was writing, thus our principal means of reconstructing prehistoric times. Prehistory, unlike history, cannot tell us about individual figures or in many cases about individual events, but rather about societies as wholes: their settlement patterns, their subsistence strategies and technologies, their social stratification, their relationship with one another as well as their interaction with the environment, both in the way the environment affected populations and how populations affected the environment. It also yields information about the movement of people, their encounters with one another, the merging or separation of cultures, the development from hunting and gathering to agriculture, and the range of social and cultural elaborations which have occurred in an area over time. It also provides information about innovations in technology and artistic styles over time and their diffusion from one society to another, as well as about migrations and conquests. There are no written records in prehistory, but in some places and some times, archaeologists uncover a graphic record in the form of petroglyphs, cave drawings and paintings, and statues and ritual objects which reveal certain aspects of the lives and beliefs of vanished peoples.

The threshold between history and prehistory, known as proto-history, occurs when a people that has not yet developed a writing system is noticed by others who have writing, and whose existence and to some extent their way of life is recorded, as in the case of Greek and Roman descriptions of the Scythians, the Celts, and early Germanic people, and in the Americas and Africa where colonial powers described some aspect of non-literate native peoples.

Archaeology is also a point of convergence where of necessity different disciplines come together. Within anthropology there is the overlap with physical anthropology in the field of bioarchaeology, the study of past human cultures through human remains recovered in archaeological sites. Ethnoarchaeology combines ethnography with archaeology by studying contemporary traditional cultures as a way of providing analogies for the study of similar prehistoric cultures, and in places where there are continuities of patterns of manufacture, trade, and construction in contemporary folk cultures, those activities can be studied for information useful in the interpretation of archaeological sites where similar artifacts are found.

Archaeologists also sometimes examine the distribution of speech communities to locate their boundaries and their movement over time as revealed by their distinctive material remains. The principle underlying this research strategy is that linguistic boundaries generally correlate with ethnic or racial groups. The difficulty here, however, is that the two do not always correspond. Also categories such as "language," "ethnic," and "racial" can be hard to delineate. Yet it is an empirical fact that linguistic and cultural boundaries do generally correspond, and that language is often a factor in maintaining cultural and demographic barriers, thus the use of linguistics for this purpose is warranted, but with caution.

Reconstructions of the vocabularies of earlier phases in the history of languages have also been analyzed to see what they have to tell us about the early cultures of which those languages were a part. And loan words allow us to track the spread of innovations that might be reflected in the archaeological record.

Social and cultural anthropology provides models from which the social organization of extinct people can be inferred from their material culture. And from the humanities the models and methods of art history are used to detect interaction between social groups as well as patterns of diffusion of styles, motifs, objects, and in some cases, the ideas they convey. Those methods and models also help in appreciating artistic conventions in ancient works as a way of gaining knowledge of the culture that produced them, of the beliefs that they may have represented, and of the aesthetic patterns that they convey.

Different branches of other sciences are also employed when using material remains in reconstructing the past, such as physical geography, geology, paleontology, paleoclimatology, paleobotony, chemistry, physics, medical technology, and sometimes astronomy for dating artifacts and establishing chronologies, or when examining human remains and for investigating settlement sites and ascertaining environmental condition of earlier times, and for analyzing and interpreting artistic objects and sites. Archaeology, like physical anthropology, is therefore a point of convergence of a number of scientific disciplines in the investigation of an established field of interest.

ÖTZI THE ICEMAN OF THE EARLY BRONZE AGE

One example of how the convergence of different sciences on a single problem can yield a wealth of detailed information of about the past is the case of an over 5,000-year-old mummy found in 1991 high in the Alps at an altitude of 3,210 meters above sea level. He was first called the Iceman, because the remains had been remarkably well preserved in glacial ice, and later Ötzi after the Ötztal where he was found.

Like the remains of Neanderthal Man, the Taung Baby, and *Homo naledi*, Ötzi was discovered by chance. Two German tourists were walking off the trail when they saw what looked like human remains partially visible in the snow. They notified the authorities who at first treated the find as a forensic matter. When the authorities suspected that the mummified remains were not those of some missing mountaineer, but rather of ancient origin, they sent the mummy to the university in Innsbruck where an archaeologist inspected it, along with accompanying artifacts which included his complete clothing, his bow and arrows, a knife, an ax, and a traveling bag which has been called Ötzi's medical kit, all well preserved in the ice for over five millennia, and now on display in the museum in Bolsano, Italy in Southern Tyrol.

Ötzi has been identified as Europe's oldest natural human mummy giving us an unprecedented look into the Copper Age in which he lived.

His age at death has been estimated at forty-five years, and by means of 3D technology his facial features have been reconstructed. A radiological examination of the bones revealed that he had in places age conditioned or strain produced degeneration. The tibia, femur, and pelvis also show that he walked long distances in hilly terrain, a degree of mobility unusual at that time, thus the possibility that he might have been a high altitude shepherd.

His teeth showed a considerable degree of deterioration from cavities, which might have been caused by a high carbohydrate diet heavy in grain. His intestinal contents showed that he had eaten two meals before his death, the last one eight hours before he died. One meal consisted of deer and herb bread, and chamois, a goat-like antelope, for the other meal. Near the body were various seeds and grains possibly in the form of bread, as well as berries and wild fruits, possibly parts of his provisions, along with some grains indicating that his death had occurred in the spring. His diet was further ascertained by an analysis of his hair to find signs of what he had eaten in the past several months.

Genetic analysis revealed that he had a high risk of artherosclerosis and lactose intolerance as well as indications he may have been the earliest known sufferer of Lyme disease. Intestinal parasites were also found and CAT scans revealed that three or four ribs had been crushed by the ice as the body lay in the snow. An analysis of a fingernail shows that he had been ill within his last three to six months, the last illness two months before his death which lasted two weeks. The epidermis was missing through the natural process that occurs when a body is mummified in ice. And an analysis of the hair showed a high level of copper and arsenic. This together with his copper ax blade indicates that he might have been involved in smelting. There were also indications of tattoos. Judging by their location they might have been part of treatment for pain. Scientists sequenced his full genome, and analyzed the blood cells, which were still intact (the oldest ever analyzed), showing his relationship to other groups in Southern Europe. It has also been suggested that he had a higher degree of Neanderthal ancestry than do modern Europeans. The examination also indicated that he had met a violent death at the hands of an enemy, for an arrow was found in his shoulder and he had suffered a head wound made by a blunt instrument.

The tools he carried are examples of how Bronze Age men were equipped and how their tools were made. Also his intact clothing gives scientists tangible evidence of what was worn and how and from what material they were made. This exhaustive study of a single individual complements information about the period which has been gathered from the excavation of a number of residence sites from that time. It also shows how the convergence of different sciences can yield a wealth of information which is applicable to any problem encountered in the archaeological study of prehistory.

PROCEDURES AND METHODS IN ARCHAEOLOGY

The process and the methods of archaeology require the skills of a researcher as well as those of an administrator, since data are gathered through careful systematic excavations often involving teams of workers and scholars whose work must be planned and supervised. Some archaeological studies begin with regional surveys; an attempt to systematically locate features of previously unknown sites, in some cases by means of aerial and satellite imagery. Site surveys on the ground include the identification of the surface signs of a site, and below the ground indications can be detected with magnetometers of minute deviations in the earth's magnetic field caused by iron objects and by some types of stone structures and even ditches and middens. Devices for measuring electric resistivity of soils are sometimes useful, as are metal detectors on battle fields where bullets, buttons, and other metal objects left from the conflict may lie underground. And in underwater sites, magnetometers and sonar are useful in locating shipwrecks or submerged settlements.

When the site is located, test trenches are sunk to produce a visible archaeological section of the site for recording. Stratigraphy enters here (the first major achievement of archaeology as it was passing from its antiquarian to its scientific stage) in which strata are identified and the artifacts or other features found in them are recorded and became a permanent part of the site record. From those samples an excavation plan is drawn up. The site is then excavated to reveal the camp or settlement which has been detected there, uncovering structures, fire pits, middens, and so forth, for further examination. Artifacts are cleaned, cataloged, and recorded as to their provenance from the site as a whole and the stratum in which they were found. They are then compared with published collections in order to classify them typologically. Other objects found in association with artifacts are such items as charcoal and animal bones; plant and pollen remains are also recorded. By means of those data the site and the artifacts can be dated.

Two kinds of chronologies can be established, relative and absolute. Relative chronologies are those in which time order is based on the relationship of artifacts or features to one another in the site, in terms of earlier or later rather than in terms of age as expressed in years. Absolute chronologies are expressed in the age of the artifacts and features themselves expressed in years. For organic matter radio carbon dating is employed. Dendrochronology, or tree ring analysis, is also important for calibrating radio carbon dates as well as in establishing relative chronologies and pollen analysis. For inorganic matter, including ceramics, thermoluminescence is used and other techniques from the field of physics, as well as archaeomagnetic dating, the interpretation of signatures in the earth's magnetic field in past times.

Information technology has provided new tools for analysis and interpretation. Computer graphics can create three dimensional models of a site and can visually reconstruct settlements, landscapes, and living conditions from the data assembled from a site. Computers are also used to build topographic models combined with astronomical calculations to ascertain whether columns or other structures are aligned with celestial events, thus indicating if they were used for astronomical ceremonies.

Two of the hallmarks of the prototypical sciences (physics and chemistry) are experimentation and replication. Although archaeology is a field science dealing with past events, experiments also sometimes play a part in understanding material that has been uncovered, how they were used, for example, and what the material indicates about ancient life. One line of experimentation is what archaeologists call working constructions of the past—or reconstruction or reconstructional archeology—the reconstruction of ancient structures using historically accurate material and technologies. Two notable examples are Buster Farm in England where an Iron Age farmstead has been replicated, and where long term experiments in prehistoric agriculture, animal husbandry, and manufacturing are carried out. Another is Lejre Farm in Denmark where a number of experiments with Bronze and Iron Age technologies and ways of life are performed. And on the English-Scottish border, part of Hadrian's Wall has been reconstructed to understand the engineering of that time.

Other experiments include the reconstruction of ancient earthworks to measure the effects of subsidence and silting in the ditches, and to get a better idea of what ancient monuments looked like. Replicas of stone tools have also been made and given to butchers, lumberjacks, and archers to use in their tasks. The observed results allow archaeologists to judge the effectiveness of ancient tools. Modern replicas of ancient artifacts have also been buried for varying lengths of time to analyze post depositional effects on them to get a better idea of the changes in the originals which have taken place over time. Other interesting experiments are attempts to replicate Damascus steel whose technique of manufacture has been lost for centuries, and graphic representations on Mochica ceramics have been used as a guide in replicating ancient Peruvian metallurgy.

SCIENTIFIC ARCHAEOLOGY

In chapter 2 we saw how, in the nineteenth century, archaeology developed from an antiquarian pursuit to a more systematic field of inquiry. Interest then was in culture history; revealing past civilizations and showing how culture changed over time as revealed by the archaeological record. Much was learned about the past from that approach, but as the field developed, a different approach emerged called at first the New

Archaeology and later processual archaeology. It began in 1958 with Gordon Willey's and Philip Philip's *Method and Theory in Archaeology*. In the 1960s, Lewis Binford and Kent Flannery argued that it was time for archaeology to become more rigorous, more scientific in its method, and more "anthropological," meaning that it should be more closely associated with other kinds of information and theories that were used in other branches of the discipline. Past societies were to be studied not just in terms of the classification of artifacts, association with other features, diffusion, and evolution, but also in terms of their technologies, living patterns, economic base, and social organization in terms of the ecology.

In 1962, Binford argued that archaeology and the rest of anthropology form an integrated whole since the structure of an entire culture is reflected in its material culture and both approaches were necessary to understand it. The integrated discipline, he wrote, strives "to *explicate* and *explain* the total range of physical and cultural similarities and differences characteristic of the entire spatial-temporal span of man's existence" (1962). He goes on to describe one way of dealing with archaeological remains which he calls a systems approach. In this approach, structural relationships between major cultural subsystems such as the social and the ideological sub-systems (as seen in the artifactual assemblages within the environment) and the changes that have taken place over time are examined to provide the basis for the description of past societies. This is what has been called cultural ecology which we examine in chapter 12 and which is illustrated in chapter 13 in the study of one culture area.

CLIMATE, EVOLUTION, AND CULTURE

We saw in the previous chapter how periods of different climatic conditions played an important role in human evolution. Such fluctuations are part of the natural dynamics of the planet, and so have continued to play a part in human life. One of the most marked aspects of ancient climates is the appearance of ice ages, periods of cooling on the earth's surface and in the atmosphere, resulting in the expansion both of mountain glacial and continental sheet ice. Those periods are also characterized by interstitial warming and contrast with longer periods of higher surface and atmospheric temperatures, during which ice sheets and glacial ice retreat. This ebb and flow has affected patterns of human life for millennia, changes which archaeologists have been able to reconstruct, showing sometimes a causal relationship between environmental conditions and, in varying degrees, the shape of cultures and the nature of population growth and at times its reduction. The fate of civilizations has also been attributed to climate change such as the fall of the Maya civilization.

This ebb and flow is recorded in sediments and tree rings as well as in the historical record. One major historical climate event was the medieval warm period from AD 952 to 1250, when the limits of agriculture were extended to higher elevations in the mountains and into higher latitudes than before, and when Vikings moved out across the North Atlantic colonizing the Faroe Islands, Iceland and Greenland. They even established for a short time a base in North America which archeologists discovered in 1960, at a site called L'anse aux Meadows in Newfoundland which is dated around AD 1000. This was followed by a cooling period during which the Eskimo population, adapted to cold conditions, expanded west while the Viking settlements in Greenland, unable to adapt, were abandoned (Diamond 2006). Another cooling period occurred in the seventeenth century which climatologists call the Little Ice Age and which historians call the General Crisis since it led to a time of instability, famine and war as seen in the historical record (Parker 1980, 2013).

Another kind of climate change, besides the long Ice Ages and warm periods, are shorter cycles of moister and wetter times. The sudden and extensive expansion of the Mongol Empire in the thirteenth century was attributed to a time of drought which pushed the pastoral peoples out of their home area as far west as Eastern Europe and southward into China to forge the largest land empire in history which covered half of the Eurasian land mass. Recently, however, tree ring data and sediments from lakes, going back to AD 900 indicate that the Mongol expansion took place not during a time of drought, but rather during a period marked by a string of wetter than usual weather and over a larger in extension than any other in the last millennium. This period, which lasted a generation, could provide richer grazing than normal and thus more fodder for horses (which made the Mongol horde mobile and so effective), thus providing greater resources for empire building. Without such conditions, it is argued that Genghis Kahn, the instigator and the leader of this imperial surge, would not have been able to mobilize forces, exercise his talent for leadership and his will to triumph in that highly successful enterprise.

This wetter period may or may not have been associated with another cycle of moister and drier times which has occurred for thousands of years but only recently identified, a periodic weather occurrence known as el Niño. This occurrence is caused by a periodic warming of the Pacific Ocean with long distance effects. Peruvians, especially fishermen, have noticed its effect off the coast of Peru around Christmastime, thus the name el Niño referring to the Christ Child. Normally the warmest ocean temperatures in the world are in the Western Pacific near Indonesia. During el Niño those warm, rain generating waters slide east creating conditions for heavy, in some places excessive, rainfall while leaving other regions dry and suffering from drought. Archaeologist Brian Fagan has examined such climatic changes and their effects on human life in his informative *Floods, Famines and Emperors: El Niño and the Fate of Civilizations* (1999), which illustrates how archaeology and paleoclimatology help explain some of the changes seen in the archaeological record.

CONCLUSION

The subject covered by physical anthropology—the biological nature and the evolution of the human species—flows smoothly into the realm of investigation of archaeologists which is the prehistory of the species after its emergence from its hominid ancestors, tracing its course into the historical period and into the ethnographic present and up to current times, the latter a field of study covered by linguistic and sociocultural anthropologists; all a seamless whole, yet each specialty within it employing different methods suitable to that part of reality it has chosen to examine, and each based on the same underlying method of inquiry, which defines them as a scientific enterprise.

TEN

Sociocultural Anthropology

Social and cultural anthropology is the largest of the four sub-disciplines in terms of the number of people working in the field, as well as the range of human activity it covers: social organization, belief systems and symbols, political and economic relations, and the relationship between populations and their physical environment. This sub-discipline is also characterized by different theoretical positions, more so than in the other three sub-disciplines. For those reasons we devote four chapters to this branch of anthropology. The first deals with definitions, methods, a historical overview, and the broad scope of the field. The second with its disparate theoretical approaches and in the third and fourth chapters we summarize the social science and the humanities approaches in sociocultural anthropology as well as briefly discussing applied anthropology.

SOCIAL AND CULTURAL ANTHROPOLOGY

The three strands of European anthropology—French, German, and British—are present in different degrees in American anthropology. British social anthropology, however, has had the greatest influence. The focus in the British tradition was on the study of still ongoing preliterate societies in the Pacific area and in Africa. With the exception of E. E. Evans-Pritchard, it was strictly synchronic. With an emphasis on social structure and organization, specific rights, duties, status and roles, and their relation to culture, British anthropology was, as self-described, comparative sociology in the Durkheimian tradition.

In America with Boas and his students, the emphasis was on culture: the shared customs, values, and beliefs that distinguish different social groups. The Boasian approach is what has been called historical particularism which holds that each culture has its own unique and particular traits and history which are not governed by universal laws as argued by the evolutionists. The traits which distinguish different cultures are explained in terms of specific cultural contexts and different unique histories. This approach fits well with the concept of diffusion, which had become the focus of German ethnology, since similarities in two cultures could be accounted for not by parallel development along some general evolutionary lines, but rather by the diffusion of traits across culture boundaries through interaction between cultures.

Boas's and his followers' interest in the historical context of culture along with their belief that all cultures matter led to concern about the loss of native cultures through assimilation or acculturation. Boasian anthropologists thus advocated salvage ethnography which became a significant part of their work. Salvage ethnography holds that it is important to gather as much information about native cultures as possible before they are lost through assimilation or acculturation, preserving each as a part of the permanent ethnographic record. Added to the recording of ongoing customs were informants' recollections of customs and beliefs that were no longer operative but could still be remembered and related.

There were also a number of written records available describing native peoples and their customs, some going back to the seventeenth century. The study of those records eventually became the field of ethnohistory, a part of both history and anthropology, with its own journal and its own professional organization, the American Society of Ethnohistory. Moreover, the study of Native American prehistory through archaeology was an established part of anthropology in its institutional setting. Thus a combination of ethnographic, ethnohistorical, and archaeological interest in a single topic, the American Indian, along with the historical particularistic perspective of the day, gave American sociocultural anthropology both a diachronic and a synchronic character.

At first American cultural anthropologists were interested in individual tribes. Ruth Benedict departed from that pattern by looking at culture as a whole. She argued that each culture is defined by a unique configuration of traits which could be recognized in individual cultures. She assumed, like others, the commonality of all human beings, but said that each culture selects only a limited number of possibilities "from the great arc" of human behavior, a selection which gives each one its specific character. In her 1934 book *Patterns of Culture* she identified the distinctive patterns of three cultures: the Zuni of the American Southwest, the Kwakiutl of the Pacific Northwest, and the Dobuans of Melanesia. She later approached Japanese culture as a whole in *Chrysanthemum and a Sword: Patterns of Japanese Culture*. Her method was called "anthropology from afar," for instead of traditional field work in which the investigator interacts with people in an ongoing community she used literature, newspapers, government propaganda, and so forth, as her source material.

Clark Wissler also looked beyond individual societies, in this case at culture areas; geographic regions identified by aggregates of cultural traits and complexes. In this way, like the *Kulturkreis* school in German ethnology, he brought together ethnology and geography around the concept of culture. He first applied his trait-complex method to the culture areas of Native North America, an approach which was later refined with more sophisticated statistics, and applied elsewhere. Alfred Kroeber's contribution to the study of culture areas appears in *Cultural and Natural Areas in North America* where he views the relationship between native cultures and geographic areas, and the historical relations found within them.

Kroeber was also interested in the wider aspects of culture and history as seen in his 1944 *Configurations of Culture Growth* where he describes long term historical patterns such as the decline and fall of major civilizations in different areas of the world such as Egypt, Mesopotamia, China, and the Islamic area, focusing in each case on such cultural features as art and philosophy, and so forth.

Those two approaches, British social and American cultural anthropology, each with its own strength and its own valuable complementary body of data, merged to form sociocultural anthropology as it operates today. There are two components in this branch of anthropology; *ethnography* which is the systematic description of a people and their culture, based on direct observation within the community, and *ethnology*, the cross-cultural comparison of ethnographic data within some theoretical framework, in order to identify individual and universal traits, as well as to reveal patterns and causal connections in an attempt to explain if possible both singularities and regularities. Ethnography is thus basically ideographic and interpretative, while ethnology is primarily *nomothetic* and where possible, explanatory.

ETHNOGRAPHY

Ethnography, like geology, is a field science since they both involve systematic observation and description with very little latitude for experimentation. The difference between ethnography and other field sciences is that ethnographers study *people,* and since people live both in the observable objective world, and in the world of meaning of their own making, the ethnographer's task is to go beyond simply observing what people do, to an attempt at understanding those meanings, or as Malinowski put it, to grasp the world from the native's point of view. One of the chief methods employed in that pursuit, and also a defining feature of sociocultural anthropology, is participant observation; the firsthand observation of behavior in its everyday (natural) setting through daily interaction with the people of the community. Another defining feature is the informal elicitation of information in everyday contexts along with formal interviews.

There will probably be some people in the community who have some mutual interests with the field worker, or people with whom the ethnographer shares some empathy, and so who are more willing to supply valuable information and native insights on their way of living and social forms than others. Thus much information comes from an unrepresentative sample. The generality of that information is checked by questions based on that information which are asked under the appropriate circumstances and by the observation of everyday activity.

There are also various levels of knowledge within a community, no matter how simple the society might be. There is the everyday, common sense kind of cultural knowledge, the kind everybody knows, but always in variation—variations from one individual to another, between the generations, and among members of different factions. This also holds for differences in what men and women know and how they perceive the same reality. Here, husband and wife teams can access both sources.

Anyone who has done any field work in rural semi-literate communities knows that almost all of the human interests and potential talents and skills we find in our own society can turn up among different individuals in those communities. Under different circumstances some of those individuals might have become lawyers, political leaders, scientists, engineers, historians, theologians, concert musicians, or painters. This fact was caught in verse by Thomas Gray while reflecting on a country churchyard at sunset in a simple village in eighteenth-century England.

> Beneath these rugged elms, that yew tree's shade
> Where heaves the earth in many a moldering heap
> Each in his narrow cell forever laid,
> The rude forefathers of the hamlet sleep.
> Perhaps in some neglected spot is laid
> Some heart once pregnant with some celestial fire;
> Hands that the rod of empire might have swayed,
> Or wake to ecstasy the living lyre.

Finding such people and talking to them about their interests and about their abilities and how they and their skills relate to the life and the activity of the community can provide a rich source of data. Another valuable source of information comes from the most highly regarded figures in the community, people who have thought about their culture and its traditions and who are widely respected for their knowledge and wisdom. They are thus a rich source of authentic information about the way ideas and customs have been considered from the inside in the most reflective manner. And when specialist knowledge is sought, the ethnographer might even become a helper to the specialist or even his apprentice, as for example those ethnographers who learned about the art of traditional Polynesian navigation.

Ethnographies can be comprehensive, as was the case of Malinowski who attempted to describe the society's culture as a whole. Others may focus on one specific topic, for example traditional Polynesian navigation (Gladwin 1970). The project may involve just one investigator, better still husband and wife teams, or even various investigators examining more deeply different aspects of the culture. Another feature of ethnography is that investigators are outsiders. This puts them in the position of novices who have to learn the culture from the inside. At the same time their outsider status may allow them to see things of which the insider may only be vaguely aware or not aware of at all due to their familiarity with it. Unlike the novice, however, the ethnographer brings to the field knowledge of how other cultures work and the variation among them, as well as theoretical concepts that help interpret data as it comes in, and to formulate new lines of inquiry as the project continues.

However, what one is looking for may affect the way the entire community is seen and depicted in the ethnography, for what one ethnographer sees may not correspond with what others see, or with data that come from other sources against which the ethnographic account can be compared. Perhaps the best known case is the study of the Mexican community Tepotzlán first studied by Robert Redfield (a study that informed his later formulation of the folk-urban continuity). Redfield saw the community as tranquil, its population contented, with almost perfect adjustments and social integration. A few years later his former student Oscar Lewis restudied the community and reported mistrust, fear, and hatred as well as malicious gossip and widespread social conflict. It seems that both aspects of the culture were there, but that the point of view of the observer was enough to emphasize one over the other.

Various objections have been raised regarding ethnographic data and its use in formulating social and cultural theories. One question is that of objectivity which brings to attention the highly personal aspect of data collecting in the field. All ethnographers strive for objectivity, yet objectivity is a relative term. We can always be more objective with physical than with social facts, but even there, as thinking and feeling individuals, there may be some variation as to what different individuals see and how they assess it. This of course is much greater when we are dealing with other people as we saw in the case cited above. The only remedy is simply to recognize the limits of objectivity and to strive to view the behavior we are observing with an open mind realizing that there is more than one way to deal with situations (more than one way to skin a cat as the old saying goes).

The ethnographer's task is to find the inner "logic" of actions and to describe it in its own terms and not those of the culture of the observer. This is called cultural relativity, a necessary perspective brought to the field by the ethnographer in an attempt to minimize distortions due to natural ethnocentric as well as theoretical or other perspectives. Early attempts at that kind of objectivity are seen in the work of Adair and Heckewelder. Distortion can also arise from too close an identification with the people with whom one is working, a danger known as "radical participation." Cushing was aware of that danger and was far ahead of his time in recognizing that it was inherent in the discipline. Yet he wrote that "the personal equation is the supremely essential thing in such researches" (1979: 17). Alfred Kroeber in the 1930 edition of the *Encyclopedia of the Social Sciences* wrote of Cushing, "what he brought to science at this juncture were the imagination and talents, and some of the quirks, of an artist" (quoted in Cushing 1979: 16).

The personal factor always plays a role in ethnography, where personality makes the job easier from some and harder for others. Recently this recognition has led to what is called reflexivity, closer attention to who has done the field work, and what manner and under which conditions it was carried out. Some of those considerations were discussed by Malinowski in the introduction to his influential ethnography *Argonauts of the Pacific* as well as by others, yet the standard format was that of an external observer who in an expository narrative presented a picture of what had been observed with no reference as to the role the ethnographer had played in the process of field work and in its presentation.

In the 1980s, however, far more attention was given to this aspect of field work as a part of the activist movement of the time, a movement felt not only in anthropology but in the social sciences in general. The problem with that approach is that it was ideologically motivated and thus a distortion in its own way. Also too much self-attention on the part of the ethnographer, regardless of ideology, can lead to too heavy an emphasis on the investigator at the expense of the description of another way of life. Still some information on the ethnographer, how he or she came to the community, the conditions under which the field work was carried out, the methods used and the difficulties encountered, much in the spirit of Malinowski's classic ethnography of the Trobriand Islanders, is useful for the ethnographic record.

Another concern is that a year spent in a community, the usual span of time for a field project, and even recurrent visits, are too short and episodic to provide data adequate for a full picture of a way of life and the conditions under which it operates. But even on the commonsense level we deal with fragmentary and fallible information to build pictures of the way things are, and all branches of science work with samples of the whole. This is especially the case for paleoanthropology where fossil evidence is minimal, fragmentary, and often discovered by chance rather than by design, and still that evidence is used to infer the course of evolution of the species.

There are a number of accounts of ethnographic field work going back to the Torres Straits expedition, as well as the work of Radcliffe-Brown and especially that of Malinowski. George Murdock also describes the ethnographic project of a team of investigators of a Pacific island in the foreword to Ward Goodenough's 1951 *Property, Kin and Community on Truck*. More recently books have been written on ethnographic field work and methods such as Michael Agar's *The Professional Stranger: An Introduction to Ethnography* (1980), James Spradley's *Participant Observer* (1980) and Karen O'Reilly's *Ethnographic Methods* (2005).

Like other field sciences, ethnography is observational. Yet, like geology and archaeology there are times when experimentation enters the picture. A notable example is the case of Polynesian navigation. In the late nineteenth and early twentieth centuries a greatly romanticized picture of the colonization of Polynesia had emerged. It praised the skills of canoe builders and navigators and envisioned organized fleets bent on a great colonizing enterprise. In 1963 Andrew Sharp challenged that view, arguing that Polynesian navigational sailing skills were limited and that the settlement of the Polynesian triangle was the result of drifting, random island sightings, and luck. Old knowledge, he said, was passed down by oral traditions which eventually allowed for more sophisticated maritime knowledge. His theory caused controversy, which led to research into the art of Polynesian maritime skills, which involved experimentation in canoe construction and methods of navigation without instruments on the open sea.

David Lewis sailed his catamaran from Tahiti to New Zealand by stellar navigation alone, and in Hawaii anthropologist Ben Finney built a forty foot craft based on traditional design and tested it by sailing and paddling it in the waters of the Hawaiian Islands. At about the same time it was discovered that traditional vessels and navigation techniques were still in use in the Caroline Islands and Micronesia. Those methods were then studied by ethnographers through participant observation and interviews and described in several publications. Also in 1971 Finney, together with Hawaiian artist Herb Kawaiwui and sailor Charles Tommy Holms, founded the Polynesian Voyaging Society dedicated to the research and the perpetuation of traditional Polynesian methods of maritime travel. The members wanted to prove that ancient Polynesians could have purposely crossed such vast stretches of ocean by traditional means. In 1975 they built a craft according to traditional specifications and in 1976 they successfully sailed it from Hawaii to Tahiti along the ancient route. The navigator was Mau Pialilug, a traditional navigator from Micronesia. Pialilug trained Nainoa Thompson who later also successfully made the trip to Tahiti.

Thus ethnographic information, together with experimentation and repetition (features of laboratory sciences) made it possible for us to better understand the navigation practices that were likely used by the ancient Polynesian navigators as they populated the islands of the Pacific, and in the process adapting to seafaring, thus strengthening our knowledge of human perseverance and the ingenuity and skill which show in that one area how people spread over the entire habitable portions of the earth, and in the process adapting to its different environments.

Mau Pialilug was far more than an informant in this line of ethnographic research. He was also a demonstrator and teacher, an important figure in preserving this particular aspect of traditional culture by fixing in the ethnographic record this last phase of a centuries old art, and thereby enhancing our knowledge and appreciation of our common world heritage. He was awarded an honorary doctoral degree by the University of Hawaii, and honored by the Smithsonian and Hawaii's Bishop Museum for his contribution to maritime history.

Ethnography deals mostly with qualitative data, but also with quantitative data in the form of the size of the population, their land holdings, and use of other resources and their economic activities. More recently quantifiable data have been used together with various interpretive models to better describe what is going on in a community and to discover and explicate the meaning and function of social institutions and patterns of behavior. Among such studies is the field work of Elspeth Ready (2015) who spent a year (July 2014 to July 2015) in a small Inuit community in the Canadian Arctic studying food security and social organization.

Despite the substantial integration of the community in the national economy, traditional subsistence strategies of hunting and food sharing are still practiced. This mixed economy also operates under different conditions of food scarcity. The question is, does the acquisition of country food (fish and game) along with food purchased in stores, present what she describes as "trickle-down" pattern from high to low food security households, with the results of community-wide food security? Or do households give when they have an abundance and expect a return at times when they are suffering scarcity, which would not result in community-wide food security? Then on the basis of this information to see how those economic arrangements are expressed in social relationships within the community.

She interviewed 110 households classifying them in terms of different degrees of food security, adapting the Household Food Security Survey Module for her questionnaire. She then used the USDA method of rounding and tabulating and the Latent Trait Model for interpreting the data. One result of the study was that models designed to measure variables in a completely commercial society do not work well in an economy in which country food, and feelings about it, are not captured in the model, and where factors such as hunting skills, motivation, and knowledge networks in acquiring country food are present.

Yet she was able to show, with the models and with social network analysis, that the trickle-down hypothesis was not confirmed, that the community as a whole was not the proper unit to understand country food sharing but that such factors as kinship ties and distance between neighbors were important in food sharing. The survey and interviews also showed that food sharing served purposes of prestige and affirmed hierarchy, which sometimes has political purposes. The use of accurate measurement thus aided in showing in a non-subjective manner just how the mixed commercial and hunting strategies operate within an Arctic community which faces periodic food insecurity, thus revealing a clear continuity of a subsistence strategy from pre-contact to modern times and its adaptation to changing conditions.

The ethnographer uses such tools and techniques as census taking, surveys, maps, and interview techniques, but in all cases the recording and managing of observed behavior is important for a useful field record, somewhat like the field record of archaeological excavations or of laboratory experiments. Along with notetaking, Malinowski also recommends that the observer keep an ethnographic diary, an account of the personal experience of the day.

Another very useful tool is the tape recorder which can record music and verbatim discourse in interviews or storytelling, or in ordinary conversation where it can capture not only grammatical and conversational forms, but also prosodic variation in speech in the study of language use in the community. Even more useful is the video recorder which can record all of that, plus entire rituals and musical performances as well as conversation interactions with eye movement, gestures, and space between interlocutors that are important for work in interactional sociolinguistics.

Film or video also offers the possibility of visual as well as verbal narrative, each with its own methods of presentation, yet both with the same goal of systematic accurate description and the recording of the facts observed. Karl Heider, one of the pioneer ethnographic film makers, tells us that the film maker must accept the demands of ethnography and think scientifically as well as cinematographically, adopting the goals of ethnography to the visual potentials of the film or video. He summarizes the differences and the similarities between the two media: the word-on-paper, and the image-on-celluloid or image-on-video, in the following way (2006: 8).

Heider also says that it is valid in presenting a visual representation of a category of behavior to edit out scenes from different footages and compose them into a single scene, as was done for the giraffe hunt in *The Hunters* and for a battle scene among New Guinea tribes in his own film, *Dead Birds*, made with Robert Gardiner. The same technique, he says, is used in the narrative presentation of events in ethnography. As an example he gives Malinowski's description of a long voyage by Trobriand

Table 10.1. Ethnographic Films

Ethnographer	Filmmaker
Begins with a theoretical problem and research	Begins with a script or an idea
Gathers data by making observations and asking questions	Shoots footage
Analyzes data	Edits footage
Writes and rewrites	
Produces a written report	Produces a film
Do not expect to either to say everything about a subject	

islanders. Heider discusses a number of attributes of the ethnographic film, stressing that the film cannot stand alone, that both film and printed material must be used together to satisfy the needs of ethnography.

In the first phase of the ethnographic film, film makers sought through voice over closed narrative to represent a film version of the material contained in written narrative, with the goal of recording, representing, and hopefully analyzing some aspect of a culture. Since then ethnographic films have become more impressionistic and far more open to the viewer's interpretation. For example, voice over narrative has been abandoned for less intrusive techniques such as interviews with native subjects in order to let them tell the story themselves in their own words. Also, reflecting the trend of reflexivity in ethnography, the ethnographer is shown in the film interacting with the people rather than standing behind the scenes and speaking in the form of a written narrative.

Another innovation is to train people within the community to make their own films about themselves in their own way as a source of information from the inside the culture, presented not in oral discourse as in the case of traditional ethnography, but in film. This use of people on the inside to collect information parallels in some ways what ethnographers have already done from time to time. An early example is the work of Boas Kwakiutl assistant George Hunt in writing an ethnography of his social group.

The first attempt at this approach in film was made in 1966 by anthropologist John Adair, from San Francisco State University, communication specialist Sol Worth from the Annenberg School of Communication at the University of Pennsylvania, and one of Worth's students Richard Chalfen. They wanted to see if it was possible to teach people in a very different culture from the one in which films emerged and where they play an important role, and to see how such a film would differ from those of Western culture. For this project they combined anthropology with communication.

They selected a group of young people among the Navajo for the project. They taught them the techniques of film making and let them produce their own films. The series appeared under the title *Navajo Film Themselves*, and was discussed in a 1977 book titled *Through Navajo Eyes* written by Adair and Worth that describes the project and its results. The book became a cornerstone text in the field that has become known as visual anthropology. Technological innovations here as elsewhere have made visual anthropology far easier than before with camcorders and digital cameras which are low cost and which produce images easy to edit on-line. A problem with this ease of film making, however, is that it can lead to simply filming what comes before the device rather than filming after proper research on what is to be filmed and after adequate field work prior to filming.

In sum, ethnographic field work is grounded in knowledge of the variations of social and cultural forms around the world. It adopts the attitude of cultural relativity with the aim of objectivity, it consists of participant observation and knowledge of the local language, and it involves the systematic recording of data on land tenure, on residence and work patterns, household composition, kinship and marriage, ritual and ceremonial behavior, occupation, communal activities and the beliefs, the values, and the aesthetic products of the community as a whole, either for a general description or as context for studies of specific aspects of the way of life of communities.

And finally, following Colin Renfrew's observation on the search and discovery of truth about the human condition (2003: 193), we see that there is something beautiful, as well as sometimes something thrilling in discovering the unknown and finally understanding a puzzle, whether in the fossil or the archaeological record or in understanding a living community through ethnography and through discovery, or in ethnology by piecing together bit by bit ethnographic information for a wider understanding, and in some cases, explanation and appreciation of the vast social and cultural variation of the species and its advancement through time.

A TYPOLOGY OF SOCIAL AND CULTURAL FORMS

All sciences must classify the range of phenomenon they study in order to examine it theoretically and explain its variation and its dynamics over time. A typology for this purpose in sociocultural anthropology arranges the various social and cultural forms encountered around the world in terms of their subsistence base, their population size and density, their social organization, form of governance, economic relations, and the organization of religious activities, from the hunter-gatherer band at one end to modern urban industrialized society at the other end of the scale.

Bands are small nomadic equalitarian communities usually consisting of between 25 and 120 individuals, the size depending on the resources exploited. The society is organized around the nuclear family with bilateral kinship as the most common form of kin relation. Subsistence is based on foraging, hunting, and in some cases fishing. Division of labor is by sex and age. There is little specialization of skill that could lead to a permanent division of labor of other kinds. Trade is minimal and reciprocity is the main form of distribution. Leadership falls situationally and temporarily to the most capable individual. Coercion, laws, and restrictions to resources are rare. There is also no priesthood. Examples of this social type are the Australian Aborigines, the hunter-gatherers of the Kalahari Desert of South Africa, and the Shoshone and the Pauite of the Great Plains in North America.

Tribal societies are based on the village or clan with larger populations than encountered on the band level. Their subsistence is based on horticulture or pastoralism, or a combination of the two where the food supply comes more from domesticated than wild sources. Land or herds are most commonly communally owned. Distribution is based on reciprocity. Trade is minor but is more developed than that of the band. Such societies are equalitarian in which power is decentralized and exercised through a network of personal or group relations. Leadership is charismatic without coercive authority to enforce decisions over group opposition. In such cases oratory is often an important aspect of life where leaders must convince others to follow their lead in warfare or other communal matters, as documented in Native North America east of the Mississippi in the eighteenth and early nineteenth centuries. The right to use force is kept within the lineage or is a claim made by the community. There is very little political or religious or economic specialization and the tribes are generally equalitarian with clans based on either matrilineal or patrilineal kin based organization. There are also pan-tribal solidarities which include voluntary societies. Examples of tribal societies are the Nuer of Africa as described by Evans-Pritchard in 1940 and the Iroquois and other eastern woodland Indians of the eighteenth century.

Chiefdoms are social groups with centralized authority in the form of a chief. Only members of a lineage of a certain clam within the wider social group may fill that office by succession. Authority is exercised with minimal of power of coercion. Membership in any clan is unilineal. There is no real class structure but individuals and clans are ranked with some ranks earned and others inherited, thus their designation as a ranked society. The chief is also the head of a system of redistribution. Tribute in the form of some valuable good is collected and returned in a ceremonial manner which serves to foster social solidarity among the kin groups and clans under the authority of the chief. A prominent example of this social form is the pre-colonial Ashante of West Africa.

State Societies rest on an agricultural base with a large densely settled population which is sharply divided into social and economic classes and with occupational specialization such as peasants, craftsmen, merchants, and political and religious specialists who are in charge of the courts, the administrative apparatus, and religious organizations. Governance is carried on within a state structure which operates within a definable territory as seen in a formal body of laws and a point of authority in a ruling class which is headed by a hereditary ruler. The ancient civilizations are examples of state societies, which eventually evolved into the complex traditional agrarian society of medieval Europe and elsewhere.

Modern Industrial Societies are characterized by an industrial manufacturing and commercial economy, with large stratified populations that are geographically and socially mobile and concentrated for the most part in large metropolitan centers and governed by highly developed state structures. This social and economic form began to take shape in Western Europe in the eighteenth century where it rapidly developed and spread around the world, eventually replacing traditional agrarian society and the few less socially and culturally elaborate societies that still existed.

THE WIDENING SCOPE OF ANTHROPOLOGICAL INQUIRY

The focus of early anthropological study was small scale, preliterate face-to-face communities represented by the few hunter-gatherer bands that could still be found and the tribal societies of Africa, the Americas, and the Pacific islands. As these social groups changed with the spread of modernization, some anthropologists looked to Europe where there still existed in some rural regions communities with traits similar to primitive communities; primitive in the sense of being primary, the kind of social arrangements and cultural forms from which more elaborated societies were derived.

In 1937, Conrad Arensberg wrote *The Irish Countryman*, an anthropological study of the peasant population of Western Ireland, in which he describes the family structure and the household and the way of life and the labor of the rural population. He also describes their deep religious commitment and the persistence, on the folk level, of beliefs and practices that go back to the Middle Ages and in some cases perhaps back to prehistoric times.

In 1954, Julian Pitt-Rivers published *The People of the Sierra*, a study of the social structure of a rural community in Andalusia in which he applied the principles and methods of social anthropology to determine the nature of European peasant society in that country. He studied the roles of the mayor, the physician, and the priest, as well as smugglers and bandits and the Gypsy population on the margins, and the still common belief in witches. His study was relevant for students in several disciplines not only in Europe but Latin America as well.

In the 1930s, Robert Redfield began a long range study of the rural population of Mexico which he called folk societies, by comparing four peasant communities in different stages of their engagement with modernization. He also drew a wider picture of traditional agrarian society in general, in terms of two interacting levels. At its base was the majority of the peasant population which he called the folk stratum and at the top was a small elite which lived in cities and which held economic and political power. The two formed what Redfield called the folk-urban continuum, and their different but interacting cultures he called the Little and the Great traditions. Among his books on those subjects are *The Primitive World and its Transformation* (1953), *The Little Community* (1955), and *Peasant Society and Culture* (1956).

Milton Singer, student of the civilization of India, was another anthropologists who in the 1950s and 1960s studied a traditional society undergoing change. He was especially interested in the modern period and conducted field interviews to ascertain how those people were, then working in the industrial sector had made the transition from village and small town life to that of large cities, and from commerce to industry. He found that they had done so without abandoning their traditional institutions and that their successful careers had been influenced by Indian culture. In 1974, he summed up his approach in *When a Great Tradition Modernizes: an Anthropological Approach to Indian Civilization*. In the 1970s he turned his interest to another, an already completely modern complex society, the United States.

Kroeber had referred to peasant societies as part societies and part cultures since unlike tribal societies they were constituents in wider social, political, and cultural systems while retaining traditional communal arrangements and ways of life. In the 1960s, there was increasing anthropological interest in those social forms, and in 1966 Eric Wolf summarized the chief features of this social and cultural type in a book titled *Peasants*. Peasants, he said, differ from farmers in that farms are for-profit enterprises producing crops for sale in the market and purchasing the factors of production from commercial establishments. Farms he said range from small family enterprises to large "factories in the field."

Farmers must respond to the market in order to make a profit. Peasants, on the other hand, do not operate business enterprises of that kind. Instead they run households which are both an economic unit and a home. Their production is geared for household consumption, for communal participation, and to meet the requirements of various kinds of outside institutions, with surpluses sold in the market economy to supply those needs. Many of the implements they use and the clothes they wear are made in the household, and labor is organized on a kinship and communal basis, unlike the employment of farm laborers in commercial agricultural undertakings. Moreover farmers are attuned to agricultural science and the use of innovations, while peasants are more tradition-bound in their technology, their practice, and their beliefs.

In the realm of religion their focus is on the concrete and the practical aspects of everyday life and on social solidarity. Their beliefs, rituals, and symbols are often magical in nature which sometimes clash with the formal religious order in which they also participate. For this reason peasants have often been described as superstitious. In religion therefore, as in the economic, social, and technical spheres, we see parallel systems operating within a larger whole. In this sense they are much like societies on the tribal level, and the same ethnographic methods developed to study tribal societies apply to peasant communities in much the same way.

The rise of cities was a result of, and a contributing factor to, the rise and the development of civilizations. Cities were also of key importance in traditional agrarian societies, and modernity is marked by their growth and expansion. Today the nature and complexity and the dynamics of cities have become a distinctive field of anthropological research. An early approach to the study of urban life was that of Lloyd Warner who was among the first anthropologists to move away from the study of primitive society to the study of cities. He brought with him the methods that had been worked out for the study of small pre-literate societies which, applied in this new venue, was called the community study approach. This led to the study of the city as a wider entity, and of city dwellers in different neighbors, occupations, social classes, and so forth, as parts of a complex whole. One method used in urban anthropology is the case history, the study of individuals and of individual families.

Oscar Lewis applied this method to the study of a poor section of Mexico City and in 1959 published *Five Families*, and in 1961 *The Children of Sanchez*, which provide insights into the lowest socio-economic level of Mexican urban life through the narratives of the informants he interviewed. Lewis claims that people on the poverty level have adapted to their marginal status with a set of attitudes that define their way of life as a distinctive cultural variety which he calls the culture of poverty. Some of the traits which characterize that culture are a feeling of inferiority, fatalism, and a low level of aspiration, as well as reliance on such social forms as matrilocal families and the use of such strategies as mutual aid which are, he says, rational adaptations to the socio-economic conditions under which they live, a cultural pattern that, he argues, is found under similar conditions around the world.

The field of urban anthropology broadened and became a definable branch of anthropology in the 1950s and 1960s. It addresses issues such as city types, urbanization and deindustrialization, and what happens to a city when its main industry departs leaving behind it abandoned neighborhoods, unemployment, and displaced workers. Other topics are how a city is planned, its dynamics, the distribution of power and influence within it, its neighborhoods, and their links to the rest of the city, migration from rural areas and the formation of new urban communities, as well as urban poverty and social issues.

German sociologist Ferdinand Tönnies saw society as consisting of two types of organization and interaction. One he called *Gemeinschaft*, or "community," defined by personal interaction, and characterized by an affective element, a sense of social solidarity, and common identity. *Gesellschaft*, or "society," on the other hand is defined by impersonal relations between persons within formal institutions. Social and cultural anthropology had traditionally studied small scale non-literate face-to-face societies and so had refined techniques for studying *Gemeinschaft* social arrangements. Sociologists, who were more concentrated on the city and urban life as a whole, operated on the *Gesellschaft* level where they devised other more suitable, methods of studying cities when seen from that perspective.

One of those methods is the research design which defines independent and dependent variables for analysis. From this perspective the city can be either the independent or the dependent variable. In some studies the city can be seen as the major factor, as in the case of migration where the migrants must adjust to urban conditions. And in other studies the city may be seen in terms of its response to external factors such as industrialization and deindustrialization. Anthropologists, when they came to urban studies, brought with them their ethnographic techniques and the holistic perspective, but also adopted methods worked out by sociologists, another case of two fields merging on a topic of common interest.

Ethnicity became a topic of interest in the 1960s as ethnic awareness in modern states became more publically visible and more politically significant. In 1969, Fredrik Barth edited a book titled *Ethnic Groups and Ethnic Boundaries* in which the contributors examined ethnicity in different parts of the world. And in 1974 John W. Cole and Eric Wolf in *The Hidden Frontier and Ethnicity in an Alpine Valley* described an ethnic community in Europe. Many studies of ethnic communities in modern cities expanded that field of study.

In the 1980s interest in nationalism began to develop, especially after the break-up of the Soviet empire, when nationalism, which many thought was on the wane, was once again on the rise in Eastern Europe. In 1983 Ernest Gellner wrote *Nations and Nationalism* in which he views the rise of the nation state and its operation from a broad anthropological perspective. Nationalism, he says, is an outcome of industrial social organization. Its operation and growth depend on several factors, innovation, occupational mobility, the effectiveness of mass media, as well as universal literacy and a comprehensive educational system all based on a shared standard language. The result is the ideal "one state one culture."

Where that ideal does not exist, as in traditional multiethnic states, the state will feel the stress of nationalistic consciousness as it emerges.

What might be thought of as an anthropological perspective, because of its attention to culture and ethnicity, is also seen in historian Benedict Anderson's important book *Imagined Communities: Reflections on the Origins and Spread of Nationalism* (1983), and in the work of sociologist Anthony B. Smith, among them *The Ethnic Origin of Nations* (1986). Anderson defines a modern nation as an imagined community, since its many members must collectively imagine what they have in common in order to form a single unit despite the often great disparities within the national territory. He sees the printing press as the main spark for that imagined communality since it made possible the confluence of emotions, characteristics, events, and history to be transmitted on a mass scale through the media and other printed material.

Smith distinguishes between ethnic categories and ethnic groups. Both are identified by common traits such as language, often religion, and other cultural similarities. The ethnic group, which he calls ethnie, is different from an ethnic category since the defining features of both are woven together in ethnies as a part of a common identity, while different groups within ethnic categories may establish their identity in different ways.

Modern nations, says Smith, are ethnies greatly enlarged, and nation states are nations (as a cultural entity) with their own separate political structure, the state. The identifying features for both ethnic group and nation are common values, a common sense of history and destiny, and common symbols and myths. Smith examines the rise of nation states and their variations in the modern world. The field defined by the perspective of Smith, Gellner, and Anderson and others provides a useful framework for anthropologists studying a number of different issues in modern complex societies.

Places where ethnic groups, cultures, languages, and states meet—borders, borderlands, and frontiers—is also an area of interdisciplinary study, a converging point where anthropologists have also been active, most notably Hastings Donnan in *Border Identity, Nation and State at International Frontiers* (1998), *Borders: Frontiers of Identity, Nation and State* (1999), Donnan and Thomas M. Wilson (eds.) *European States at Their Borderlands*, and Bjorn Thomassen's "Border Studies in Europe: Symbolic and Political Boundaries, Anthropological Perspectives" as well as the role of boundaries, borderlands, and frontiers in language (Custred 2011).

Regional Studies have also attracted the attention of anthropologists, such as Latin America, Africa and Asia. In the 1970s, with increased interest in ethnicity, borders and cities in complex society, Europe emerged as a category of anthropological study, a field summarized by Jeremy Boissevain in "Towards an Anthropology of Europe" in Boissevain's and John Friedl's 1975 book *Beyond the Community: Social Processes in Europe,* an area of study which has been worked out in various ways in a number of other studies.

Anthropological research has also expanded into other areas in complex modern societies such as the study of occupations, from cocktail waitresses and firehouses to the professions. A notable example of the latter is medical anthropology. Anthropologists also work as employees and consultants for corporations which are concerned with the nature and the operation of their "corporate cultures" while others study governmental bureaucracies from the inside. In such settings anthropologists bring to the study of formal institutions, their specialized method and concepts, those developed in the small scale preliterate societies where they started, namely the methods of ethnography and the important concept of culture.

And finally there is the old question of migration, a phenomenon that has characterized *Homo sapiens* and its forerunners and human history ever since, and which has been studied by anthropologists in different parts of the world along with their colleagues in other disciplines, a processes which is fully underway on a massive scale worldwide under the new postmodern conditions of globalization. Jeffrey Cohen and Ibrahim Sirkeci (2011) have summarized this current phase of migration as a field of study within anthropology in their book *Cultures of Migration: the Global Nature of Contemporary Migration.*

ELEVEN

The Theoretical Diversity of Sociocultural Anthropology

There is greater theoretical disparity in sociocultural than in physical anthropology, a reflection of the difference between the social and the physical sciences in general, since theories in the physical sciences rest on a consensus of empirically verifiable phenomena in the material world, a consensus lacking in the social sciences which operates in the more complex human sphere. We divide here major theoretic perspectives in sociocultural anthropology into three categories simply for the purpose of presentation, implying no consensus among the theoretical positions in either category. The first deals with social structure and organization, the second with the psychological dimension of culture as well as belief systems, symbols and symbolic interaction, and the third with the relationship between populations and their physical environment.

SOCIAL STRUCTURE AND SOCIAL ORGANIZATION

From the middle of the nineteenth and into the first years of the twentieth centuries, culture and society were viewed in terms of what in anthropology would later be called unilinear evolution, since following biological evolution, culture was seen as evolving in a single line from simpler to ever more complex forms. This theory was replaced in the 1920s with another analogy from biology which was called structural functionalism, or simply functionalism. This way of understanding society viewed the whole of society as a complex system seen in terms of the function of its constituent parts in the form of norms, customs, traditions, and social institutions all working together, like organs in a body, to promote stability and solidarity.

This view goes back to August Comte who saw the body as the closest analogy to the way society is structured and operates, and biology as the most compatible model for the social sciences. This too was the model followed by Emilie Durkheim in the formulation of sociology based on what he saw as the organic unity of society. It also formed the basis of Marcel Mauss's anthropology and the theoretical perspective of the British social anthropologists.

Two different versions of functionalism merged in social anthropology. One was that of Bronislaw Malinowski who saw the nuclear family as the basic building blocks of society with the clan as the result and with myth operating as the charter for social institutions, all operating together as functioning parts of society as a whole. He also saw language not as a separable abstract structure, "philology," as he called that perspective, but rather as a mode of social action. He took the analogy further into biology when he argued that social practices could be explained by their ability to satisfy social needs, what he called imperatives.

A. R. Radcliffe-Brown added the word "structure" to his version of the theory, which he called structural functionalism. He held that most of those societies which lacked a centralized authority, the kind studied by anthropologists, were based on the association of corporate groups, and unlike Malinowski's imperatives, he argued that the fundamental units of society were processes of humans living together and interacting with one another. Social structure, he said, was both a model and the reality of social relations. The task of the anthropologist is to identify the fundamental units of society and the function of social practices within total structure of the society.

Those views of society and culture were eventually replaced by other models, yet the insights and the ethnographic data they provide form basic parts of contemporary sociocultural anthropology. Moreover, the idea of evolution would come up again when viewing culture and society in the long term, and functionalism would once again appear in the form of cultural ecology.

CULTURE, MIND, AND SYMBOLS

The mind has long played a role in the study of culture as seen in Bastien's psychic unity of mankind and *Elementargedanken*, in Levy-Bruhl's *Primitive Thought*, and in Franz Boas's *Mind of Primitive Man*. Wilhelm Wundt, a pioneer in modern psychology, also believed that the capacities of human beings, whether civilized or primitive, were equivalent. He also employed ethnological data in his explanations. The two fields, however, diverged, with psychology developing as an ahistorical field of study that regarded psychological processes in general, while anthropologists examined different cultural variations around the world within an evolutionary frame of reference. Freudian psychoanalysis eventually brought them together.

This was seen in a prominent movement at the center of American anthropology in the 1930s and through the 1950s, which was influenced directly or indirectly by Freudian psychoanalysis with the prominence of topics such as human development and enculturation, adult personalities and the meaning and function of myths, symbols, and ritual. Psychoanalysts were also interested in ethnological data and in anthropological perspectives such as Sigmund Freud and C. G. Jung, Erik Erikson, Geza Roheim, and others. Eric Fromm also employed anthropological methods in his work.

Psychoanalytic concepts were popular throughout the social sciences and the humanities during the first two thirds of the twentieth century, and as part of that trend anthropologists in general drew heavily on psychoanalysis. Some like Melford Spiro examined ways in which cultural symbols and social systems provided defense mechanisms or in some way alleviated psychological conflicts. Spiro also explored psychoanalytical concepts cross-culturally such as the Oedipus complex. Some such as George Devereux examined mental illness cross-culturally and others like Abram Kardiner studied ways in which social processes could affect different sectors of the population as well as mental health. Also influenced by psychoanalysis were those involved in the popular line of research known as culture and personality.

There was no doctrinal core among those studies but they all examined the relationship between the psychological interactions of individuals and the cultures in which they were raised, an interaction that plays a part in shaping a person's thoughts, emotions, and norms in the socialization process to conform with the requirements of his or her society. Different patterns of socialization in different cultures thus result in different personality types. This approach was made popular by Margret Meade in her 1928 ethnography, *Growing Up in Samoa*. The idea that cultures can be described as personalities was developed by Ruth Benedict and Irving Hallowell among others. The study of modal personality was pursued most prominently by John and Beatrice Whiting, Cora Dubois, Florence and Clyde Kluckhohn, and sociologist Alex Inkeles. It was from that perspective that they also studied national character. The study of myths, rituals, and cultural symbols was also influenced by psychoanalytic concepts in which cultural variations were regarded as manifestations of underlying mental processes, and in Jungian terms universal archetypes.

As interest in Freud and psychotherapy faded, interest in the cognitive aspect of the mind became prominent in both psychology and in linguistics. In the late 1950s and in the 1960s this interest crystallized into cognitive anthropology which has been defined as "The study of the workings of the human mind in social life" (Casson 1981: 7). Anthony F.

C. Wallace was among the first to develop this line of thinking, developing the idea of the mazeway, mental maps that connect personalities with their cultures and through which individuals interact with their cultural environments, conduct their daily life, and cope with stress. This concept became an explanatory element in his theory of revitalization movements.

Ward Goodenough, within the general framework of cognitive science, proposed a cognitive definition of language as "whatever one has to know in order to communicate with its speakers as adequately as they do with each other and in a manner which they will accept as corresponding to their own" (1964: 37). He also defines culture in precisely the same terms.

> Culture is whatever one has to know or believe in order to operate in a manner acceptable to its members, and so in a role that they accept for themselves. [It is] what people have to learn as distinct from their biological heritage, must consist of the end product of learning: knowledge, in a most general, if relative, sense of the term. By this definition, we note that culture does not consist of things, people, behavior, or emotions. It is rather an organization of these things. It is the form of things people have in mind, their model for perceiving, relating and interpreting them. As such, the things people say and do, their social arrangements and events, are products or byproducts of their culture of their culture as they apply it to the task of perceiving and dealing with their circumstances (1964: 36).

Cognitive anthropologists view culture as a system of public symbols employed to convey meanings between minds. At the same time culture is a private cognitive system whose ultimate locus is in the minds of individuals. People have cognitive systems for organizing and making sense out of external factors (Wallace's mazeway) which affect their lives such as natural objects and events and human behavior and emotions. Each person's cognitive map is somewhat different due to different factors such as their position in society and personal and shared experience, yet there is a nexus of shared categories and rules which guide them and which form the basis for coherent social interaction. Other formal systems besides language, such as natural logic, are also involved in this process, thus the assumption that mental data can be analyzed by means of formal descriptive and analytic systems such as mathematics and logic. Within this conceptual framework, some ethnographers examined the different ways that different languages conceived of common semantic domains as seen through the vocabularies of their language (chapter 12).

In the late 1950s and 1960s, a number of anthropologists focused on certain semantic domains within culture, such as color, kinship, and plant and animal taxonomies, as revealed in the lexicons of their languages. They devised methods of studying them known as ethnoscience or ethnosemantics employing methods such as query-response frames and componential analysis which analyzed the structure of terminological subsets within a language and which realize in language underlying shared cultural categories, methods which have proven useful in ethnography especially in dealing with kinship systems (chapter twelve). In 1969 Brent Berlin and Paul Kay examined color terminologies of the world and found that there were seven stages in the complexity of color terminologies from those at the lowest level of sociocultural elaboration, where languages have only two basic color terms to those, like English, which have eleven basic color terms. The model was subsequently refined but the relationship between language and culture and levels of social and cultural complexity was made evident by the data.

The refinement of the concept of natural category (Rosch) in cognitive psychology, the development of cultural models drawing on schema theory, as well as studies in metaphor (Lakoff and Johnson) led to an elaboration of the cognitive approach to culture. Anthropologists, such as A. Kimbal Romney, Roy D'Andrade, and Dan Sperber, have examined the action of individuals in groups, their manner of thinking and how models are internalized, shared, and distributed from this perspective. Cognitive anthropologists have also worked with social psychologists in terms of cultural universals as well as in the use of particular models of categorization and inference.

Related to cognitive anthropology, through its focus on language and underlying structures, is structural anthropology as formulated by Claude Levi-Strauss, a strain of thought that represents a French influence on American anthropology. Levi-Strauss views cultures as shared symbolic systems which are the cumulative creation of the mind. The investigator thus seeks to discover in the structure of social organization, myth, art, language, the mental principles that generate them. Although external conditions—adjustments to the environment, subsistence opportunities, and provide the raw material for the mind—constrains cultural and to a certain extent shape it, the world of thought is free to roam, resulting in diverse but formally similar patterns, like a kaleidoscope of different elements that can take on a number of different configurations.

Levi-Strauss sees the structure of culture as a model which could be perceived within social interactions. His aim was to examine the way the same domain is structured in the cultures of different societies, most clearly in the domain of kinship, in order to identify the universal properties of the structure and the processes of the mind of which those structures are a part. His theory is that the mind does not passively receive imprints of collective representations, but rather that it consists of innate structures which are "pre-programmed" and are not affected by interaction with the environment, a concept of structure which differs from that of the Radcliffe-Brown school which regards structure as the result of social interaction.

Levi-Strauss argues that those mental structures impose themselves on representations of reality in the form of social institutions, and become more selected and more specialized in different cultures. Structures can thus be seen, for example, in the way kinship systems are put together forming a "logic" (where logic is defined as "the necessary relations among things") which can be explicated by structural analysis in much the same way that linguistic structure can be explicated by studying speech. He approaches myth from this perspective by looking not at the contents of myths as a reflection of specific cultural realities, or as a charter for social institutions or as the realization of some deeper psychological entities such as archetypes, but rather in terms of the way they are put together cross-culturally constituted, thus representing in their structure the way the mind works in organizing beliefs and ideas. He calls that form of organization "mythologic" which he describes in terms of binary oppositions taking the study of phonology as his model. Although his system for inferring abstract properties of the mind is intuitive it has stimulated much research in cultural anthropology.

A long-standing question deals with different modes of thought in different times and in different cultures. One such way of seeing things and acting on them is magic, a topic which anthropologists have long dealt with for no other reason than that it is found everywhere, and was a part of Western Civilization up and into Early Modern Times.

Levy-Bruhl looked more broadly at primitive mentality which he differentiated from that of modern Man in several ways. The mind of primitive man, he said, was pre-logical and mythopeotic. It makes no distinction between the real and the supernatural in a union which he called "mystic participation." The Western mind on the other hand is rational. It confronts contradictions and speculates using logic. Despite its shortcomings, especially Levy-Bruhl's formulation of his observations within a unilinear evolutionary framework, he did isolate some features which represent reality, and which are useful today when thinking about different cognitive styles.

James Frazer, using ethnographic and historical material available at the time, described the thinking underlying it in terms of a logic based on what he thought was a belief, sometimes unspoken but in other traditions spelled out to some degree, in a pattern of sympathies and antipathies among things in the world as well as the idea of contiguity (things that are parts of a whole) on a hidden or occult level which provides a sense of causality and thus a rationale for magical practices. Marcel Mauss looked carefully at the kinds of thinking that underly magic in general, and Evans-Pritchard described one particular system, based on systematic field work, showing how beliefs operated within the Azande society of Africa.

Robin Horton working in Africa has examined the differences between the magical mode of thought, which he describes as a closed system, and the open system of science (Horton, 1970). And Stanley Tambiah has taken a long look at the relationship between magic, science, and religion, a tripartite division not found in other cultures, or in medieval Europe. He examines the origin of those ways of thinking and feeling in the Western tradition and how they developed into what we have today, and how those different strands operate in their own way in other cultures.

The question of modes of thought and external factors has also interested investigators outside the field of anthropology but which in various ways have influenced similar studies within the discipline. For example, classicist Eric Havelock in *Preface to Plato* (1963) examined the question of why Plato attacked poets. Havelock argues that it was because Greece at that time was passing from a non-literate to a literate culture where different forms of discourse were necessary for the storing of collective experience necessary for cultural stability. In non-literate societies such experience is stored in the form of poetic discourse since the poetic process makes it easier to memorize such knowledge and to hand it down, a form of discourse which necessarily presents content as a series of specific images with emotional identification within a continuous narrative. Platonic thought was one aspect of an increasingly literate society which required a vocabulary and syntax which was more abstract and precise in which to preserve knowledge and to think more analytically.

Anthropologist Jack Goody further developed that line of thinking and eventually developed the idea that there are oral and literate cultures, each characterized by different ways of thinking, and that the medium—word of mouth versus writing—plays a central role in the difference between them. Classicist Walter Ong in *The Technologizing of the Word* (1982), further developed the idea by describing primary, secondary, and residual orality in different societies over time. In sum, the theory holds that writing promotes abstract ideas while concrete concepts dominate in oral cultures, promoting reasoning and new ways of categorizing. Literacy also allows for different versions of a story to be recorded and then compared, stimulating the processes which advances thinking about the past, moving it from legend to history (see Olson 1994).

The thesis that literacy influences thinking was examined in the 1920s and 1930s well before that line of thinking began with the work of Havelock, Goody, and Ong. Lev Vygotsky, a Russian psychologist who was deeply affected by Marx's and Engel's cultural materialism, was interested in the potential of social science to bridge the gap between the social classes. Mental processes, he thought, were affected by "signs," including language. Just as changing tools according to work (tilling with tractor instead of the hoe, for example) requires different patterns of behavior, so too changing sign systems (from oral to written language) would restructure mental activities.

In the 1930s, another Russian psychologist, Alexander Luria, tested those ideas in a peasant community in Central Asia that was undergoing rapid change. In efforts to modernize the region, the government introduced land reforms, new tools as well as schooling. Luria chose three test groups from among the peasants, one that was non-literate, another from the same village which had gone through brief courses in literacy, and another which had participated in short teacher-training courses. A variety of tests were administered which dealt with perception, word association, concept classification, and reasoning.

The results of those tests, along with information from interviews, showed consistent differences in performance among the three groups which Luria interpreted, within Vygotsky's theoretical framework. That group representing the most traditional and isolated population, with neither literacy nor schooling, responded to the tasks in a concrete and context-bound way, guided by perceptual and functional attributes of things. Most schooled groups on the other hand tended toward a more abstract approach and responded to the conceptual and logical relationship among things. The minimal literacy group fell in between. Similar experiments were done in America some forty years later when researchers compared preliterate preschool children with school children of various ages, as well as with educated adults, to test the theory that literacy yields a unique form of logical competence. Their evidence showed the lack of such competence in preliterate school children while school children and educated adults possessed it (Scribner and Cole 1981).

The question then was what is the deciding factor in determining cognitive change in regard to literacy? Was it that learning to read and write alone was enough, or were other factors involved in the process of schooling that might account for the major shaping force in the difference, of which literacy was only one factor among others? To that end Sylvia Scribner and Michel Cole conducted an ethnographic study among the West African Vai using various psychological tests which showed that specific claims made about the effects of literacy alone on cognition cannot be supported. They conclude that cultures with symbolic technologies such as writing "push cognitive growth better, earlier and longer than others" but that literacy was not the sole deciding factor (1981: 11).

What all this shows is that there are differences in cognitive configurations which are shaped by external factors, which brings us back to Levy-Bruhl's dichotomy between primitive and modern thought. C. R. Hallpike in his 1979 book *The Foundations of Primitive Thought* begins by defending the term "primitive" which he uses here not in the sense of superior/inferior, but rather in the sense of a "primary" or basic form or foundation upon which other more elaborate forms are derived. Based on an examination of a number of ethnographic examples and psychological tests within the framework of Piaget's developmental psychology (how children develop cognitively as they grow), he argues that the thought of those individuals in small-scale preliterate (primitive) societies is less powerful than that of modern scientific thought. He uses Piaget's framework of developmental psychology to demonstrate this. Everyone everywhere, he says, has the basic level of thought as described by Piaget. It is, however, submerged in modern society by accretions such as education, literacy, and technological thinking. Thus on the most basic level of life where only basic thinking is required, that is as far as it goes. In more complex social, cultural, and technological environments, cognitive abilities will increase accordingly. Everyone everywhere, he says, has preoperative thought no matter how old they are. It is, however, submerged in modern society by accretions such as education, literacy, and technological thinking. Thus on the most basic level of life where only basic thinking is required, that is as far as it goes. In more complex social, cultural, and technological environments cognitive abilities will increase accordingly.

This position quite naturally attracted quite a bit of criticism. Some have said that the comparison of stages of the individuals' psychological development is inappropriate for understand the characteristics of whole communities. Others have said that no society would be viable if all its individuals operated on that basic cognitive level, and still others find fault with Piaget's ideal type, and some critics see it as a version of the crude Victorian evolutionary model. Yet one can also read Hallpike's argument as the description of different modes of thought adapting to different conditions, and his argument as an attempt to justify the use of psychological data for ethnological purposes. What interests us here, however, is the different ways anthropologists have addressed differences in the way people in different cultures perceive, think about, and make sense of the world in which they live.

Clifford Geertz also sees culture as a symbolic system associated with the mind. He comes at it however from a different perspective, though, one illuminated by Weber's interpretive sociology, and the humanistic tradition, for he said that philosophy and literary studies could contribute to the socials sciences. He cites Weber's observation that man is an animal suspended in web of significance he has spun himself. Culture, says Geetz, is that web and its analysis should not be that of "an experimental science in search of law but an interpretive one," one which is essentially semiotic, in search of meaning. It is explication I am after, he says, "constructing social expression on their surface enigmatical" (1973: 5). He then goes on to characterize ethnography in much the same way as Dilthey and the neo-Kantians describe the method of *verstehen*, as opposed to *erklären*. In this way Geertz was acting in a scientific manner, in the broader sense of science, whether he recognizes it as such or not.

Geertz also believes that the formalistic approach of the cognitivist is spurious, for meaning, he says, is not in people's heads, but rather that symbols and meaning are shared by participants in social interaction. He describes its study as semiotic and thus a matter of interpretation which he says becomes "thick description" (a concept borrowed from the philosopher Gilbert Ryle) in which the ethnographer interprets what actors are doing, not on the surface level, but in terms of intentions and the layers of structure that comprise the meaning system of the culture in which they live. It is through those "piled up structures of inference and implications" through which the "ethnographer is continually picking his way." Geertz says that "to look at the symbolic dimension of social action—art, religion, ideology, science, law, morality, common sense—is not to turn away from the existential dilemmas of life for some empyrean realm of de-emotionalized forms; it is to plunge into their midst" (1973: 30).

Symbols and their relationship to culture was also a topic of Eric Wolf who argues that there are key symbols in a society which inform meaning in a culture. As an example he gives the national symbol of the Virgin of Guadalupe in Mexico. Victor Turner has also developed what is called symbolic or interpretive anthropology. He introduced the concept of social drama to account for the symbolism of conflict and crisis resolution among the Ndemba villagers of Central Africa. He also explored the nature and function of ritual among the Ndemba in *The Forest of Symbols* (1966) and formulated a wider theory in *Dramas, Fields and Metaphors* (1974) and elsewhere. Another figure in symbolic anthropology was Mary Douglas who wrote *Purity and Danger*, an analysis of the concept of ritual purity and impurity in different societies and in different times which is considered one of the principal books in social anthropology, as well as her *Natural Symbols* (1970) which deals with an individual's role in groups.

David Schneider holds the position that symbolic meaning exists as a separate abstract level independent of situated social interactions, and that cultural symbols should be studied separately from social life. He defines culture as "a shared system of symbols and meaning: a system of categories and units and the ways in which these are designated or conceptualized" (1984: 196). Anthropology then, he says, "is the study of particular cultures *The first task of anthropology,* prerequisite to all the others, *is to understand and formulate the symbols and meanings and their configuration that a particular culture consists of*" (1984: 196, emphasis mine). Cultural analysis he says should be directed toward the system of symbols embedded in the normative system from which they can be abstracted.

By symbols and meaning he means the basic premises the culture posits for life, how they integrate the rules and classifications of the various domains of experience and how they all relate to one another and to "the world in which man sees himself living in" (1972: 38). This, he says, is the prior task before tracing the interconnectedness between the cultural, social, and psychological life of groups or the actions of individuals. He further says that symbolic anthropology is very close, if not identical with interpretive anthropology.

CULTURE AND THE PHYSICAL ENVIRONMENT

People are social and cultural creatures, but they also live in the material world from which they receive their sustenance and in which they live out their daily lives, sometimes at their peril. Their culture and society are thus connected to that broader material reality, a fact which became more apparent in anthropology during the 1950s and early 1960s.

Leslie White, contrary to trends of his time, championed the nineteenth century evolutionist approach, and adopted a deterministic materialistic view of cultural evolution. He saw culture not in terms of its variability, but rather as the totality of human activities everywhere that was continually evolving. There were three levels in the process: the technological level, the level of social organization, and the ideological level. The most important of the three is the technological level which determines the shape of the other two. Technology, he says, is the means by which a society harnesses energy in order to assure its survival. The more efficiently this is done the more advanced is the society in evolutionary terms from the simplest societies that harness energy through hunting and gathering, to the harnessing of energy through the domestication of plants and animals to industrialization with its use of fossil fuels that make possible a larger more elaborate society and culture. His materialist view is clearly stated in a quote from his 1959 *The Evolution of Culture: The Development of Civilization to the Fall of Rome,* "man is an animal species, and consequently culture as a whole is dependent on the material, mechanical means of adjustment to the environment."

In *Theory of Cultural Evolution: The Methodology of Multilinear Evolution* (1955), Julian Steward also looked at the broad sweep of human prehistory and its evolutionary nature, yet with a more particularistic approach, stressing the need for investigation of particular evolutionary sequences and the description of the different courses of evolution in different cultures as well as what they all had in common. His was thus a view of evolution which emphasized similar evolutionary stages and parallel developments. One example of the latter is the similar social and cultural transformations of hunter-gatherer societies of the North American Great Plains and on the South American pampas when the horse was introduced by the Spanish at the same time. Another is the consequence of the appearance of irrigation under similar conditions but separated from one another in time and space.

Steward, however, confined his approach to parallel sequences of development and neglected to apply his approach to the development of unique sequences or to consider the general, universal process. Marshall Sahlins and Elman Service in *Evolution and Culture* (1960) took that additional step by viewing general evolution as successive levels of new all-round evolution produced by the interplay over time of many cultures and sequences. In this framework, and based on empirical data, we see that some cultures make a strategic contribution to general evolution while others may stagnate. Some, like the Eskimos, become increasingly narrow in their culture as they adapt to an ever more hostile environment while others may eventually fail after handing their ideas, their skills, and their social patterns off to others, as in ancient irrigation societies of the Near East and the Peruvian coast.

Sahlins and Service, like Steward, emphasized the particular as well as the general (the ideographic and the nomothetic) in their view of human cultural evolution. In looking back over the development of this "new American evolutionism" Eric Wolf says that it may represent "the achievement of a new degree of scientific maturity" in anthropology (1974: 31). It also represents a degree of integration between two closely related sub-disciplines of anthropology, sociocultural anthropology, and archaeology, which have come together in the spirit of the discipline's holistic approach. This integration features the role of the environment in the life and the development of cultures and societies, something that had previously been ignored.

In the late 1960s theoretical approaches emerged in anthropology that put the material aspect of culture in first place, harkening back to the past, to Marx and Engels, Malthus and social Darwinism, and incorporating the evolutionary views of Leslie White and Julian Steward. This was the theory of cultural materialism first introduced by Marvin Harris in 1968 in his book *The Rise of Anthropological Theory*. This approach was more fully developed in his *Cultural Materialism* which appeared in 1979 with a second revised edition in 2001. One example of his materialistic functionalist approach is seen in an article on the sacred cow of India written in 1963, when Harris described that ancient religiously based custom as an example of economic and cultural rationality.

He eventually drifted away from Marx, rejecting Marx's Hegelian dialect and calling the Marxist relationship between infrastructure and superstructure mystifying and unscientific, later saying that Marx was not his chief inspiration but rather positivistic behavioral psychology. In fact, Harris represents the strongest example of positivist materialism in anthropology. His research plan and his work have been criticized as reductionist, influenced by materialist philosophy and positivistic psychology with simple schemes, simple interpretations with preordained conclusions.

TWELVE

Anthropology as Social Science

Besides its theoretical disparities, sociocultural anthropology ranges across the spectrum of social and cultural activities, thus dividing into sub-fields which overlap other disciplines in both the social science and the humanities. In this chapter we examine those subfields that lie within the realm of social science.

At first, British social anthropologists, with their emphasis on kinship, kin groups, status, role, and social structure, thought of themselves as comparative sociologists in the Durkheimian tradition, a position which neatly fit into the more culturally oriented American cultural anthropology thus defining a clear sociological aspect of contemporary anthropology. As its scope broadened to include the study of groups and communities within complex social systems, the methods of social anthropology overlapped with those of sociology so that, as discussed in the previous chapter, the ethnographic methods of studying social organization in small scale communities complemented the methods of sociology which cast a wider net.

Another area of traditional anthropological interest is the study of another of those universal concerns of human nature, the persistent questions, who are, where did we come from, what is our place in the cosmos and where is it all going? Levi-Strauss in *The Savage Mind* writes about the way prehistoric technology came about and developed. It was through the process of bricolage, he said, the constant tinkering with things, and ordinary informal trial and error to see what works best. So too, he said, did myths emerge from images and ideas to answer the questions "who, what, to what end?"

Analogy plays a leading role in this process in common sense, as in philosophy and science and as in myth. You take what you know and use it to make sense out of what you don't know what you feel is there—

what somehow lies behind the effects of things you know for sure such as disease, accident, misfortune, the "best laid plans" that go awry when they shouldn't. From that foundation emerged, in the minds of Men, concepts of spiritual essences of things, and spiritual forces and beings which, once phrased in those intelligible terms, can be comprehended, dealt with, brought under control, placated, venerated, and lived with for better or for worse. The crystallization of such means of explanation, accommodation, and control into bodies of shared belief and practice gave rise to magic and religion.

Here we once again encounter a semantic problem, in this case the difference between magic and religion. Religion is usually defined as a system of thought and belief which is shared by a group (a community of faith and when organized as a social institution, a church) that gives its members an object of devotion, a code of conduct, and a conceptual framework in which to relate to the universe, and which acknowledges some transcendence as some part of the system of belief. Magic is the means of influencing or compelling spiritual forces or entities for the purpose of achieving the practical results. An apparently universal belief that appears in varying forms, which can be considered a part of magic, is belief in witchcraft, the use of magic in some form by an ordinary member of the community for the harm of others (Marwick 1970). The best description of magic in this sense, as a system of practical explanation and control, is found in the ethnographic work of Evans-Pritchard (1937).

One differentiating feature between magic and religion is that magic is an individual affair, or one involving a client and a practitioner, while religion is a community affair. Another is that magic is pragmatic, conceptually elaborated only as far as necessary for rationalizing its practice, whereas religion involves higher speculation and more elaborated formulation. The demarkation between magic and religion actually makes no sense on the most basic social level as seen, for example, in hunter-gatherer and other such small-scale societies, where the only specialist is a shaman or some other individual who is believed to have certain power, and where the belief system is a part of the folk culture of the group. As society develops and as occupational specialization becomes pervasive, religious specialists appear in the form of a priesthood, and the belief system and its myths become more formalized and more carefully worked out and elaborated. All this is the domain of the anthropology of religion.

Economic Anthropology studies the exchange of goods and services within a society, an activity described to some degree in most ethnographies. Special attention was brought to exchange in 1922 by Malinowski in the *Argonauts of the Western Pacific* with his description of a pattern of exchange known as the Kula Ring Trobriand islanders, who made voyage with the purpose of exchanging bracelets and necklaces. Malinowski asked "why should men risk their lives and limb to travel across huge expanses of oceans to give away worthless trinkets?" His answer was that the exchange enhanced social status and prestige and thus had a value for the participants, but one different from the exchange of resources.

Marcel Mauss, basing his study on ethnographic descriptions available at the time, describes non-economic exchanges in different places in his 1925 book *The Gift*. Mauss makes a distinction between commodities, simple alienable items which are bought and sold, from objects which embody other values such as reputations, histories, and identities of various kinds. Giving those items does not involve economic advantage but rather the establishment of patterns of mutual obligation and reciprocity between giver and receiver which are important in interpersonal, and from his perspective, intergroup solidarity.

Drawing on ethnographic examples, and especially *The Gift*, economist Karl Polanyi wrote in 1944 *The Great Transformation* and in 1957 *Trade and Markets in the Early Empires* in which he argued that market exchanges represent just one form of exchange, a form characteristic of Western industrialized societies, but that other forms of exchange based on reciprocity and redistribution and embedded in kinship, religion, and political life are characteristic of small scale societies and early empires. The discipline of economics, he says, is based on the study of market exchanges in western societies and is not universal. He thus draws a distinction between what he calls the formalist and the substantist approaches to the study of economics, a distinction that started a debate that would affect the anthropological study of patterns of exchange in the ensuing years.

Albert Schrauwers, in examining economic relations in Indonesia, says that prestation, the kind described by Mauss, among noble lineages, preserves landed estate identified with particular kin groups and it serves to maintain the place of those groups in a ranked society. Also in Indonesia, Clifford Geertz identified what he called a "dual economy" and James C. Scott on different "moral economies," different spheres of exchange which emerge in societies which were becoming integrated into a market system. In both cases Geertz and Scott hypothesize a continuing, culturally ordered traditional sphere of exchange which persists along with the market economy. Another example of the relationship between traditional indigenous and market exchanges was described among the Tiv of West Africa by Paul Bohannan where three spheres of exchange operate with certain goods exchanged in each sphere, each with its own form of money.

Marshall Sahlins, in *Stone Age Economics* (1972), hypothesized that there are three aspects of reciprocity in the exchange of goods in small-scale societies. One is the gift or generalized reciprocity which involves the exchange of goods and services without keeping track of the exact value of the items exchanged, in many cases with the expectation that in time a balance will result. Another is symmetrical or balanced reciprocity where one makes a gift with the expectation of some tangible return in terms of a specific time, place, and amount. The third is market, or negative reciprocity—the exchange of goods and services in which each party intends to profit in some way at the expense of others.

Generalized reciprocity he says operates within kin groups on no matter what level they are found (a feature of *Gemeinschft* social relations). The more distant the exchange partners are from one another, the more imbalanced or negative the exchange becomes (a feature of *Gesellschaft* social arrangements). Thus different dynamics of exchange operate in different ways and in different variation depending on the simplicity and the complexity of the society in which they are transacted.

Other anthropologists such as Paul Bohannan, Raymond Firth, George Dalton, and Harold Schneider have argued that neoclassical economic models could be applied anywhere since they are based on universal principles, if properly adjusted to the situation. One aspect of this approach is the rational actor which is an ideal type in the economics model whereby one assesses the costs and benefits of an exchange against a standard of value and needs and desires. This works ideally in the market as well as in various forms of reciprocity. For example, exchanges in the Kula Ring are rational despite the risks involved in voyages and even though the commodity value of the items exchanged are, in Malinowski's words, "trinkets."

When examining the different kinds of commodity exchanges we see the practice of barter in stateless societies and elsewhere, where one item is exchanged for another of what is considered commensurate value. There are also various kinds of money involved in trade, some are for special purposes, items of symbolic value, that are used to facilitate trade in different spheres such as shell money and wampum. And then there is universal money, which facilitates trade of all kinds with abstract measures of value. This kind of money provides a general standard of direct and deferred payment, can be used to pay debts and taxes, and which allows wealth to be preserved over time.

This brings us from small-scale and larger stateless societies to modern local, national, and global markets with its banking and finance systems, the bond and stock market and corporate capitalist culture. This, too, is a realm of economic anthropology. What, for example, are the particular ways that the free market is involved in local political and cultural life and how do they operate in different countries? The realities and the culture of Islamic countries are not the same as those of the West, and China operates within its own cultural context, despite the fact that they all interact in an intimate, mutually influencing way in global economics. Here the insights, methods, and theoretical formulations of economic anthropology can help better understand this aspect of exchange on a complex and dynamic world-wide scale.

Political anthropology is the study of political organization and political structures in different cultures and in different times. We find political organization in some form even on the most basic hunter-gatherer level of society as seen in the ways that groups attend to the necessary tasks of maintaining internal order and managing external relations. On the level of small-scale societies, such as those on the hunter-gatherer level, internal order is a maintained through custom and interpersonal social rewards and sanctions. For external relations and for communal activities, where decisions must be made, leadership roles fall to those recognized as the most qualified for the task at hand. Leadership in personalized equalitarian polities are therefore much like the kind of leadership one encounters in modern communities on the local or neighborhood level.

Larger, "tribal" societies are more complex, consisting of different lineages of kin-based groups. In many cases those groups are ranked, with some regarded as commoners while others are ranked higher, having achieved the status of a nobility. The political organization of such ranked societies has been called that of chiefdoms, a form of political organization based on kinship where formal leadership is monopolized by a senior member of a family or a "house." Thus a leadership role is ascribed in a high ranked kin group within the society, whereas on the equalitarian level the status is achieved.

A more elaborate form of ranked polity is feudalism that in this context can be defined as the relationship between commoners and the nobility, and among noble families themselves in a hierarchical relationship to one another, based on customary rights and mutual obligations. The Bunyoro of East Africa are an example of this socio-political type. A "Great King" ruled over a territory which was divided into a number of local regions each ruled by a local chief, some with only a few villages, others with many more. On the village level the chief would gather grain, beer, and cattle which were then offered to the chief above him who would take a share of the tribute and pass the majority of the wealth up the hierarchy. In return the chiefs redistributed the wealth on ceremonial days and provided aid to the population, which had produced it in the first place in times of need.

At this point it is useful to stop and consider the terms "authority" and "power" which are in some cases intertwined but which are not synonymous. Authority is the right and the power to determine, adjudicate, or settle disputes in some other recognized manner, the right to control or determine or direct someone or something. Power is the ability to control or influence the behavior of people, the ability to act, and to accomplish something. Power may be backed by the authority of law or tradition which gives the ruler an official right to control and determine and is thus regarded as legitimate by the social system. Power always implies coercion, the use of force or intimidation to make someone or a group act in an involuntary manner. For example, a policeman can force compliance both by the authority vested in him by the state, backed up with the ability to command through coercion. Personalized equalitarian societies, on one end of the continuum, rely more on the authority of leaders based on custom and confidence in their ability for regulating affairs, while at the other end states rely on authority and coercion in order to govern. The study of political organization thus involves attention to the flow and confluence of power and authority within the polity.

With the increase in population density of sedentary agrarian societies, this form of kin based personal political organization develops into a more formal and impersonal governing structure, the state, which consists of permanent governing institutions managed by a bureaucracy and staffed by impersonal officials. The state thus forms an entity outside and above the networks of personal associations and obligations that constitute the society. The king rules now through the state which plans and executes enterprises of its own initiative in accordance with visions of the ruling elite, and are implemented through state action backed by coercion. Ranked polities were financed by means of tribute, goods given to a ruler for protection or as an acknowledgment of submission or allegiance. The state transformed tribute to taxes, specific amounts of goods, later money, paid at specific times, to support the government and its enterprises. Tax evasion or the refusal to pay was punishable by law through the exercise of force.

States evolved in only a few regions of the world and spread outward from there. The original locations were Egypt, Mesopotamia, the Indus Valley in China, and Mesoamerica and the Central Andes in the Americas. Once developed, states began to expand, creating the empires that figure so prominently in the history of civilization, empires which rise, thrive, decline, and fall, leaving after them to a certain extent cultural unity over a wide area, followed by a period of political fragmentation and cultural diversity. One prominent example of many is the peak and the fall of the Roman Empire.

At its peak the Roman imperial state ruled an empire that stretched from Mesopotamia to the Atlantic, and from southern Morocco to the English-Scottish border. It eventually split into two parts, the Western Empire and that of the east ruled from Constantinople. The eastern part developed its own distinct Byzantine culture while the West fell into decay from which emerged in the eighth century AD a system of feudalism from which later dynastic states emerged which in turn developed into the modern nation state.

Anthropologists have always been interested in the political aspect of society. We see this in the work of Lewis Henry Morgan, Sir Henry Maine, and others. In British social anthropology political anthropology goes back to 1940 with the publication of African *Political Systems* edited by Meyer Fortes and E. E. Evans-Pritchard where they wrote that "a scientific study of institutions must be inductive and comparative and aim solely at establishing and explaining uniformities formed among them and their interdependences with features of social organization" (1940: 4).

This approach was designed to classify societies into discrete categories for comparison from which generalizations could be made without reference to their historical contexts.

The emphasis there was the structural-functionalism of Radcliffe Brown which assumed that societies were well defined entities which sought equilibrium and social order. Edmund Leach, however, saw the necessity to understand how societies change over time rather than remaining static in equilibrium, a perspective he developed in 1954 in *Political Systems of Highland Burma*. Max Gluckman and the Manchester School of social anthropology approached society in a different way, by focusing on social process and analyzing structures and systems based on relative stability in which conflict was seen to maintain stability. The first full fledged expression of political anthropology as a sub-discipline came in 1960 with Victor Turner's and Marc Schwartz's (eds.) *Political Anthropology*.

In America, some anthropologists, Morton Fried, Elman Service, and Eleanor Leacock, extended the tradition of Morgan, from whom Marx and Engels had drawn some of their ideas, by following a Marxist approach to political organization. In that way they sought to understand the origins and development of inequality in human society. In the 1960s, the European colonial systems were being dismantled which provided new situations for anthropologists to study. It was at that time that the process approach was developed, which stresses the role of agents in the political process (Bailey 1969; Barth 1969).

From 1860 to the 1960s, political anthropology had been concerned with politics in stateless societies. In the 1960s, however, attention was increasingly turned to complex modern societies and with them a whole host of new fields of inquiry in which states and bureaucracies enter into ethnographic descriptions, one of which was the study of peasant societies and in studies of rural communities in Latin America, Algeria, and India. This was also the time when attention was increasingly turned to Europe with studies of communities in Spain, Greece, Turkey, and Italy, all of which had to take into account the state and the market as important factors. Ethnicity and nationalism also became of interest and with it the examination of the nation state and its emphasis on the correspondence of political and ethnolinguistic borders which in turn involved the relationship between ethnic and linguistic minority within the borders of nation states.

Following other trends in anthropology, anthropologists began to bring the ethnographic approach to the study of state institutions from the inside the way they were studying from the inside the real (as opposed to the ideal) operation and culture of corporations and other modern organizations. Economic and political anthropology, therefore, have increased in scope and grown in maturity thus expanding our understanding of basic parts of the human experience from the beginning to the present.

Another part of the human sphere which anthropologists have examined (much less however than in other domains) is one of the essential characteristic not only of humans but of all mammals; namely, the play element in the lives not only of children but of adults as well, to the point that Jan Huizinga has called the human species *Homo Ludens*, Man the Player (Huizinga 1955). An aspect of that feature is entertainment, whether it is woven into more practical activities or pursued solely for amusement, from the folk level to the emergence of specialists such as singers and authors, and organized events such as festivals and theatrical and musical performances.

A part of entertainment is sports, the organization of games and teams and competitions which in contemporary times have become not only big business for managers and players, but also rallying points for group identification and solidarity. The Olympic Games is one important example, another is international soccer. Long-standing symbolism is embedded in the game of soccer as seen for example in the unofficial "Christmas Truce" of 1914 on the Western Front before the First World War grew into an all-consuming fury, when British and German troops paused not only to sing carols but also to play soccer with one another between the lines.

Today, in an increasingly secular Europe with its de-emphasis on the nation state, the iconography, chants, and celebrations of soccer events have overtones of the secular religion of nationalism. This is especially the case of Germany, partly because of the many successes of its teams, but also because soccer is a way of showing the flag and singing the national anthem and otherwise expressing patriotism without evoking suspicion colored by the country's past. Children's play, organized sports, and entertainment thus offer a field of study for anthropologists in one of the characteristic features of the species.

Ecology is the scientific study of the relationship of living things with their physical environment, and ecosystems are plant and animal communities interacting with one another and the environment, concepts that form the basis of cultural ecology, another sub-field within sociocultural anthropology. Human communities were long considered as separate entities to be studied in terms of their social organization and culture. The tendency to adopt an ecological approach, says Robert Netting, which began in the late 1950s and the early 1960s, was not a conscious adherence to a new theoretical approach or the testing of hypotheses, but rather, among some anthropologists emerged from dissatisfaction with formulations that were felt to be too vague along with increased interest in culture change.

The terms of reference in which the new interest in ecology developed was broadly outlined by Julian Steward when he defined cultural ecology as "the adaptive processes by which the nature of society and an unpredictable number of features of culture are affected by the basic adjustment through which man utilizes a given environment" (Netting 1986: 6). The excitement that an ecological approach brought to anthropology was not to overturn established theories but rather to put old ideas in a more comprehensive context. That context had always been latent in anthropology through descriptions of subsistence activities, a part of every traditional ethnographic description. Now it was manifest in the form of an inclusive scientific framework that corresponds to reality. Functionalism was extended beyond the social sphere and structural arrangements were now seen to have adaptive values in a more inclusive framework, taking in traditional social and cultural anthropological knowledge and methods together with the concerns and methods of biologically based ecology. In this regard, Netting points out that "the ability of another set of facts to help make sense of what was already known by pointing out further order and meaning is, after all, at the heart of scientific endeavor" (Netting 1986: 6).

Themes in current cultural ecology are the continued interplay between changing environments and the learned behavior of human populations within them. This involves decisions arrived at by individuals and the adaptive strategies devised by them and by groups within this interaction, as well as the evolutionary processes of which these forms constitute a part. Ideational or symbolic aspects of social life are seen in this context as ways of behaving and believing that validate forms of social identities and satisfy aesthetic needs within this broader field of interaction. Thus the ecological perspective in anthropology includes the political culture and the organization of group life, its forms of economic organization and exchange, as well as the ideational and symbolic aspects of culture.

Clifford Geertz's 1963 study of Javanese land tenure, *Agricultural Involution,* was influenced by the work of Julian Steward. Gregory Bateson drew on cultural ecology for his contribution to general systems theory and Marshall Sahlins and Marvin Harris, who develcped his own version of materialism, were also influenced by this approach. Since hunting was the subsistence base of early humanity, any theory of the nature of man and society requires information on that way of life. Cultural ecology offered a way to learn more about that stage in social and cultural evolution. A symposium was held on that topic, published in 1968 by Lee and Devore, titled *Man the Hunter,* which was one of the major works in that area.

Another high point in the study of culture from the ecological perspective was Roy Rappaport's 1967 study of the Tsembaga Maring of the New Guinea highlands titled *Pigs for the Ancestors,* which has been described as an instant classic in the field, combining social and symbolic anthropology with biological ecology. Rappaport examined the intimate connections between the adaptation of the Maring to both their physical and their social environments. He ends with an analysis of the aesthetic language they use to conceptualize their world view: a holistic approach including the components of the social and cultural systems with the physical setting in which they live.

In the previous chapter and in the above sections, we saw how the different social science subfields of anthropology have dealt with some of the basic aspects of human behavior, and in that way extending our comprehension of societies and cultures in their wide range of variation, and thus giving us a keener understanding and appreciation of the human species.

APPLIED ANTHROPOLOGY

The distinction between the theoretical and the applied sciences and their relationship to one another is well known. The best examples are from engineering which applies the principles of physics for the construction of bridges, roads, and so forth, and of the practice of medicine which employs biochemistry and physiology for the diagnosis and the treatment of diseases. In the case of medicine the relationship between research and clinical application is close and mutually supportive.

Although theory has dominated anthropology, practice has also long been one of its features. The term "practical anthropology" was used in 1860 by John Hunt the founder of the anthropological society of London, and Edward Tylor considered anthropology a "policy science" and a means of improving the human condition. The concept of applied anthropology as a separate branch of the discipline first appeared in a diploma program at Oxford University in 1906. Also, social anthropology, starting in 1908, was used by the British in the administration of their foreign possessions, an application mentioned by Malinowski in his description of anthropology as a science. Anthropology was later used for that purpose by other European colonial powers.

In the United States, anthropologists were active as liaisons between Indian tribes and the Bureau of Land Management. Most notable in this respect was the work of John Collier, Commissioner of Indian Affairs in the administration of Franklin D. Roosevelt, who applied anthropological understanding of native cultures and values to help reconstitute tribal government and to reformulate school programs. Today anthropologists are active in behalf of tribes, the government, and business enterprises in transactions between the parties.

In 1941, the Society for Applied Anthropology was founded and its *Journal of Applied Anthropology* was first published. The name of the journal was changed in 1949 to *Human Organization*. The stated purpose of the society was "The promotion of scientific investigation of the principles controlling the relations of human beings to one another, and encouragement of the wide application of these principles to practical problems." Their main fields of interest were mental health, the study of industrial organization, and economic development. The specialist publication *Journal of Economic Development* appeared in 1951.

During the 1970s the practice of anthropology expanded, and in 1978 the journal *Practicing Anthropology* was published to bridge the gap between academic and practicing anthropology. In 1983 the National Association of Practicing Anthropology was founded with members representing all four fields of anthropology, although archaeologists also have their own organization.

The methods of applied sociocultural anthropology are for the most part the same empirical methods that were first developed for the study of small-scale preliterate societies along with other techniques in the social sciences: interviews and participant observation, together with focus groups, social surveys, and statistical sampling as well as textual analysis and archival research. One of its principal features is the concept that organizations or associations of various kinds develop their own cultures and that this aspect of behavior is important in understanding the nature and the operation of formal organizations, and that understanding a corporate culture in turn can be used to improve its operation. Clients of applied anthropologists include governments, development agencies, the health care professions, public service agencies, businesses, and tribal and ethnic associations.

Archaeologists have been employed by businesses for environmental impact studies which are required before construction of infrastructure, housing, or commercial developments can begin. Archaeology also forms the backbone of Cultural Resource Management defined as the management of cultural resources such as art and heritage sites, a field which also incorporates what is known as Cultural Heritage Management. In that capacity archaeologists are employed to recover and restore sites for public display and interpretation, as well as for recreational and educational purposes.

Physical anthropologists have also long served as forensic advisors assisting in the identification of human remains by ascertaining the age, sex, and racial affiliation of skeletal remains. Other aspects of forensic anthropology are the use of oral traditions and language as evidence in court cases involving human remains. A well known example is the case of Kennewick Man, the ten-thousand-year-old skeleton discovered on the banks of the Columbia River in Washington state whose ancestry became an issue in a court case in 2003 (www.friendsofpast.com) in which questions of oral traditions, linguistics, and archaeology were also addressed (Simic and Custred 2003, Godard and Shipley 2003).

Another legal area in which anthropologists have been involved is that of Indian land claims. The earliest case in which anthropologists were involved was that of the *Indians of California v. the United States*. Alfred Kroeber served as the plaintiff's director of research. Owen Stewart from the University of Colorado was his associate director and Ralph Beals was the director of research for the federal government, and several of Kroeber's students served as witnesses. Kroeber's impact on Indian Claims Commission may have established the manner in which expert witnesses presented their testimony before the court. The early interest and the expansion of applied anthropology since the 1970s indicate a healthy, useful, and thriving field both of theory and practical application.

THIRTEEN

Anthropology and the Humanities

From the start, sociocultural anthropologists were interested in the social organization of human groups as well as in what people within societies must collectively know and believe in order to function as communities. Eventually this interest expanded into the examination of the relationship of communities with their physical environment, and of the interaction of communities with one another in ever more complex levels of social and cultural organization.

Another aspect of the human experience is artistic expression as revealed in objects, actions, words, sounds, and events of various kinds. The methods used here are basically the same as the methods employed in attempting to understand human social organization and its relationship to the environment: careful observation, systematic classification, a search for patterns (in this case in themes and styles) in an attempt to "understand" in the sense both of interpretive social science (how artistic works and systems of thought relate to the sociocultural whole), and in terms of what those objects, events, images, imaginings, and performances mean to the people who produce and use and appreciate them. Such studies involve both the nomothetic and the ideographic aspects of scholarly (scientific) inquiry.

THE WILL-TO-FORM, AESTHETICS, AND APPRECIATION

Viennese art historian Alois Riegel (1858–1905) saw in the abundance of art in every age what he called *Kunstwollen*, the "will-to-form," a fundamental aesthetic urge which lies at the core of the motives of artists. In accordance with that urge, he says, each age generates its own concepts of truth and beauty which are manifested in the form of art. Otherwise the age and the culture could not express itself. Riegel argued for the autonomy of that urge and sought to demonstrate it and its fluidity through time by using the methods of philology and historical criticism.

Besides the aesthetic urge to create, there is also an aesthetic response by the public to what has been created; a quality seen in the object which, as Raymond Firth puts it, provokes "some combination of cognitive and emotional elements"; a quality which has been described since classical times, and which has, over the centuries, been crystallized in the concept of "beauty." Firth, Citing R. G. Collingwood, says that the root of aesthetic sensibility lies "in imagination, exemplified in the pursuit and enjoyment of beauty" (1992: 18). Beauty in this particular sense refers to the emotional value given to an object, or to a form or to an action which gives pleasure and satisfaction to the mind—the sense of quality. The meaning of beauty, however, differs from culture to culture, from age to age, and even among individuals, yet despite the variations, the concept and the feeling are still there. The ways of judging the aesthetic value of something also differ, but the urge to create and the basic sensations involved in contemplating what has been created seem to be universal.

Aesthetics, however, is not restricted to objects of art, for human actions can also be judged aesthetically and enjoyed as such, for example the way one is dressed and the graceful movements of individuals in everyday life or in athletic activities. Utilitarian objects can also have aesthetic value such as the Golden Gate Bridge, or the design and operation of machinery, as when railroad buffs appreciate the lines of a steam locomotive and its life-like operations and when they hear the thrilling sound of its whistle. Practical pursuits can also be perceived and felt aesthetically as for example when hunters in a hunter-gatherer society properly and successfully accomplish the joint action of detecting, pursuing, and bringing down their prey. Abstract mental constructs can also be aesthetically appreciated such as elegant mathematical constructs and well formulated scientific theories. For example, Nobel laureate Max Born once said of Einstein's theory of general relativity, before it was empirically borne out, that it appealed to him "like a work of art" (Bartusiak 2015: x).

Aesthetic emotion can also be stimulated by natural forms such as sunsets and landscapes or by the motion of animals, like the flight of a flock of swallows or a gazelle in motion, and by the very grandeur of nature itself as expressed by Alexander von Humboldt when he described "the universality of Man's enjoyment of nature" which is encountered "in every region of the globe, in every stage of intellectual culture, the same sources of enjoyment are alike vouchsafed to Man" (1849: 3).

Subrahmanyan Chandrasekhar, on receiving the 1983 Nobel Prize in physics, said of black holes that those massive objects in the cosmos, long doubted and now confirmed, are "the most perfect macroscopic objects there are in the universe" and in this regard that "beauty is the splendor of truth." Black holes, that mysterious cosmic phenomenon, says Marcia

Bartusiak, offers "all that physicists yearned for in a theoretical outcome: both simplicity and beauty" (2015: xi). And finally Frank Wilczek, 2004 Nobel laureate in physics, in his book *A Beautiful Question*, sees the symmetry and order in nature in terms of its inherent beauty, for we see in nature literally the same equations that govern atoms, light, sound, and musical instruments. Nature, he says, seems to embody beauty and elegance (Wilczek 2015).

Another aspect of beauty and aesthetic evaluation is appreciation, the recognition and enjoyment of what is considered the good qualities of something, as well as a full understanding and a more accurate perception of the object and a true estimation and assessment of its value. Such assessments are sometimes made explicit on the folk level, which can be elicited and described by the ethnographer, and on a more sophisticated level, expressions of appreciation merge in the work of the critic and the more scientifically oriented work of the scholar (see chapter 5).

THE ANTIQUITY OF ART AND THE ETHNOGRAPHIC PRESENT

Nothing indicates more strikingly the transition from hominid ancestral forms to modern Man than the extensive paintings found throughout Eurasia and Australia dating back to as early as 50,000 years. However the earliest concrete expression of the conscious use of design was found in South Africa in the form of a small block of engraved red ochre which goes back 75,000 years ago (Ochre has been used in Africa for millennia for food preservation and for body decoration). The etchings were carefully carved patterns of triangular, overlapping markings. No one knows what those markings meant to those who carved them, but the artificiality of design is clear. Subsequent searches in the region uncovered similar objects going back 100,000 years.

The most spectacular prehistoric art was found in the Lascaux and Chauvet Caves in France and Altamira in Spain. Those Paleolithic paintings depict hunting scenes produced in many cases with a grace and style that pleases the eye today. Just what role they played in Paleolithic culture (perhaps magical or religious) we do not know, yet we do know that their creators felt some urge to produce them, an urge which has been revealed in religious art, or in art simply for the purposes of enjoyment, ever since. Attempts at interpreting those charcoal drawings and paintings is of course based on speculation, but some features are highly suggestive, and conjectures as to how they were used, and perhaps even how they were aesthetically evaluated, are worth considering, if for no other reason than to show their artistic potential regardless of the original intention of the artists or the way they were received by their contemporaries.

One question asked of those paintings is, why were they painted in the deep recesses of caves? Magico-religious reasons immediately come to mind, and if that was the motive and the use (which seems reasonable when seen in the light of symbols, ritual, magic, and religion in societies all over the world and through time), then speculation as to how they may have operated in such contexts are worth considering. First of all, it is possible that the pitch black darkness of those remote, sometimes hard to reach, chambers played a part in the process, perhaps as a backdrop for the interplay of light and dark and the flickering effect of fire light from stone lamps.

One observer, Chip Walters, says that bumps and crevices on the rock wall of the cavern, on which the figures were painted, might have suggest natural shapes to the imagination, the way passing clouds do to children. In the cave at Altamira, for example, a sense of life and dimension is given to a bison by incorporating humps and bulges of the rock into the painted image. There is a panel in the Chauvet Cave depicting four horse heads drawn over subtle curves and folds on a surface of receding rock which accentuates the figures' foreheads and snouts. Walters describes the way the figures change when seen from different perspectives. One view presents a perfect profile, but seen from another angle the horses' necks and noses seem to strain, as if running away from the viewer. Cave lions seem to emerge from the wall of another chamber accentuating the haunch in one lion's back and shoulders as it stalks its unseen prey. In depictions such as these, both the medium and the lines and color are used for effect so that it seems that some of the animals were here all along just waiting to be revealed by the artist (Walters 2015).

Some images also have a kinetic effect. Archaeologist and film maker Marc Azéma thinks that the paintings and drawings were man's earliest cinema; superimposed images combined with flickering lamp light in the profound darkness of the cave to create an illusion that the paintings were moving, something like flip-book animation. Jean Clottles was the first scientist to see the paintings in Chauvet. He believes that by their effect they may have been intended to be experienced much the way we experience movies, or theater drama or perhaps more appropriately, religious ceremonies. If that were the case those paintings were components in a show, probably based on a magico-religious ritual, which served to lift one out of the real world and to bind the audience together in a shared, meaningful experience. Walters says that "even now you can still feel the power of that shadow show as you walk through the chambers" and in the silence of the cave one can imagine "its rhythm of ancient music, the beat of the dance, as a storyteller casts the light of a torch upon floating images and enthralls the audience with a tale" (Walters 2015: 57).

In modern times, Edmund Carpenter (1973: 59–75) describes the way traditional Eskimo artists (now called Inuit) went about carving objects from ivory, bone, and antlers. The process, he says, was far more than just about the object itself. It was also an act of seeing and expressing life's value, and a ritual discovery by which images and patterns are revealed through the process of carving. A carver holds the ivory in his hand, turning it over and looking at it from different angles asking "who are you? Who hides in there?" then he whisper, for example, "a seal." The carver rarely sets out to carve a seal, or anything else, at least consciously. Instead he picks up the ivory, examines it, and finds within it its hidden form. If the form is not immediately apparent he carves aimlessly until he sees the inner form, humming to himself or chanting as he works. Then the hidden seal emerges. It was always there. The carver did not create it. He just released it. He helped it come forth.

George Swinton understands Inuit aesthetics in much the same way. He describes the aesthetic not just as a concern with pleasing or with the beautiful but as "sensual cognition" which involves the well-made rather than the beautiful. He cites the Inuit word for a print or a carving which he translates as "making a diminutive likeness or model." This, he says, is the major principle of Inuit aesthetics; that ultimately the quality resides in the successful achievement of giving tangible reality (i.e., form) to the subject matter. The major concern here is the desire to be as "real" as possible and that goal is achieved through sensual or tactile means. To be real, however, does not entail the exact reproduction of the subject. Instead sensuous cognition is found in the simplification of the form, whereby the essential qualities of the subject are expressed. Textures can be smoother or rougher than in reality depending on need, and the body action of the subject "becomes translated into form through increased tensions, bulging volumes, [and] intensified proliferation of shapes" (Swinton 1978: 71–86).

The mass of people in small-scale non-Western societies feels the need for beauty in their lives although that concept is thought of in different ways. Moreover their sense of the aesthetic seems to be as intense on the whole as is that of people living in complex civilizations even though their range of artistic expression is limited. The difference is that conditions of civilized life, with its greater occupational diversity and greater wealth, make it possible for those who are artistically inclined to spend their time in artistic pursuits, producing objects considered of value, and thus creating a higher demand with greater sophistication, variation, and volume. On all levels the will-to-form and aesthetic feelings and evaluation, and thus art, are present and thus a field of research for cultural anthropologists.

ART AND ANTHROPOLOGY

At first most anthropologists approached art from an evolutionary or diffusionist perspective, or they studied it in terms of types associated with culture areas or with tribes or with schools. Since the 1960s, however, attention has increasingly turned to the study of the meaning and function of art within society, combining sociocultural anthropology and archaeology, and increasingly involving an interdisciplinary perspective in which anthropology, art history, and aesthetics overlap and mutually support one another, another case of the convergence of sub-disciplines and disciplines focused on a common point of interest.

The distinction between arts and crafts and between the high arts and popular and commercial art are features of modern Western society, and not of other societies. The task of the anthropologists studying arts in other societies is therefore to discover the distinctions made in those cultures as well as their varying ideas of the aesthetic. The role of the artist also differs in different cultures, but in all cases artists are individuals with certain aptitudes and dispositions. Their significance as artists, however, differs from culture to culture and from one age to another. In modern Western civilization they enjoy a high status but elsewhere they may be no more than any other craftsmen or simply people who produce such objects as a pastime.

Anthropologist Alan Merriam approaches the study of art with the methods of observation and directed interviews with artists and with the public. In such studies questions are asked about the significance that art has within the society in which it is produced. Also, do objects of art tell a story, what feelings do they evoke, and what aesthetic emotions do they stimulate? And what is the use of art objects: as decoration, as an expressive part of ritual and religious in the form of paintings and statues, or are they purely *objects d'art* created and admired for their aesthetic quality alone? And how do such objects function in society? Does their possession confer or signify prestige and power, or are they a part of a ritual or do they function as emblems of a wider group and thus contribute to social solidarity? Also, what aesthetic values do they express, sometimes overtly express, when evaluating not only obects but events and practical activities (Shelton 1992, Cootes 1992).

Another approach is that of Albert Gell, who sees art as a form of instrumental action. Making things, he says, means creating objects by which the thoughts and actions of others can be influenced. Thus art objects embody complex extensionalities which mediate social agencies. In Gell's theory, persons and art merge. From that perspective he explores the psychology of patterns of perception, a theory taken from linguistics and philosophy.

THE GRAPHIC AND PLASTIC ARTS

Art objects are either missing or scarce among hunting and gathering peoples. Boas attributes this to the fact that hunters and gatherers live a nomadic life with little material possessions beyond what they need to live, partly because they must travel light in the pursuit of their way of life. Their *Kunstwollen* is expressed less in objects than in song and story-telling. Sedentary populations on the other hand have the space and the time as well as better technical means for producing decorated utensils such as cloth and pottery and ritual objects which can be used and which may be regarded aesthetically. The more stratified and occupationally diversified a society becomes, the greater the variation, the sophistication, and the quantity of artistic objects.

Art as an activity is dynamic, with external factors affecting it in several ways. A good example of this process is described by Nelson Graburn in "Inuit Native Art: Miracle of the Twentieth Century," in which he shows how, under changing social and economic conditions, traditional Inuit art became a commercial enterprise and eventually an "art" in the Western sense of the term. The only material art of the pre-contact and early contact period was carvings on bone, ivory, and antlers for games and toys, such as dolls for little girls. Everything else they produced was based on practical considerations as defined by their life as hunters in an Arctic environment. After contact, however, they became involved in the fur trade and thus in association with passing mariners who traded various items for their carvings.

Farther west, Alaskan Eskimos had found a market for their carving in the Northwestern United States, and in the early 1950s the production of carvings became a commercial enterprise there. A similar development appeared in Canada as fox fur brought the Inuit little and new ways of making a living were needed. In order to strengthen the local economy the Canadian government gave a grant to the Inuit in 1949 to develop their handicrafts. They also sent James Houston to encourage the development of native art for the market. The production of souvenirs was also encouraged by the Hudson Bay Company. Inuit carvers were shown pictures of what products would sell, and the native artists accordingly produced what was in demand. The carvings depicted animals, people, and mythic themes, and exhibited what was considered a "primitive look," a trait which was valued in the carvings and later in Inuit art in general.

Not only was there a financial motive for that enterprise, but an appreciation of skill and workmanship was also involved; for example, appreciation of the particular way an object was made. Authenticity was also valued by both the artist and the consumer. Inuit art soon expanded from carvings on ivory, antlers, and bone to stone sculpture and then to print making.

Drawings were produced by artists, and block cutters would then cut blocks on which the drawings were copied for printing. Printed fabrics were also woven for sale. Later stencils were made and the copper plates, later blocks, were carved directly. There was a lot of trial and error involved and some of the shops had to close. Later co-opts were formed which were aided by subsidies. As the business expanded, the art kept developing, becoming more aesthetically elaborate, more stylistically diverse, and more individualized and characterized by greater innovation and creativity.

Some producers of those objects came to see themselves as artists in the western sense. Some of them now live in cities and some even attend art school and participate in the art business. What used to be an indoor part time activity is now an occupation of its own. What once involved a trade for a pack of cigarettes now involves payments of thousands of dollars. And as opportunities increase, art has become less important to their economy and more "art" in the Western sense. Commercial success allowed the process of elaboration to advance without a clash between the new and the old, for there was no consensus that their thousands of years of art was undergoing change. Also, for urban Inuit native art is a tie with their tradition and thus a part of their identity. What we see here is an example of the adaptation of a traditional art form to new conditions and a response to new opportunities which lies at the heart of the continual human experience of change, adjustment, and elaboration.

VERBAL ART

Although the simplest societies had little or no material objects of art, they did create verbal art in the form of song and narrative. In the case of hunter-gatherers in the Kalahari Desert of South Africa, for example, narrative and poetic language played a role in passing on knowledge, especially of a specialist kind while at the same time performing an aesthetic and an entertaining function. Only later does poetry (the specific use of meter, tropes, and rhythms distinct from ordinary prose) emerge verbal poetry from song. Oral literature, or verbal art, in song and story is thus perhaps the oldest form of the artistic urge that later proliferated in other media and genres as societies grew and became more diversified and as cultures became more elaborate.

Oral storytelling and song are by definition performances which involve the interaction of performer and audience. In understanding such performances the participants as well as the story are taken into account as well as setting and purpose. Sometimes audience participation is a part of the event, and in some cases more than one performer is involved, as in the Eskimo poetic duals and Polynesian song competition. Oratory is another form of verbal art which is usually defined as "the art of addressing an audience within political, legal, ceremonial, or religious settings" and among anthropologists and linguists as "the skillful performance of particular speech genres within ceremonial or religious magico-religious settings, such as curing session, initiations, weddings, and funerals" (Duranti 1992: 154). Oratory in all its cultural contexts can be a force for social mobilization as seen in the oratory of Native Americans in urging peace or war. It also fosters social solidarity in times of stress as in speeches of consolation and defiance, or reaffirming values and goals when they are endangered, the Gettysburg Address being one of the prime examples of that genre.

Oratory differs from other ways of speaking in the same community by its concentration on metaphor, proverb, and its use of parallelism and repetition which gives it a poetic ring and a special appeal. With the introduction of writing and the growth of literacy, a new turn was taken in verbal art in the development of different genres such as the short story, the novella, and the novel in prose narrative, and different forms of poetry. The essentials of verbal art, however, remain the same.

Oral narratives in the form of myth and legend, as well as ordinary narratives of personal experience and storytelling for entertainment, can also reveal much about the way of life of a people as seen in its recurring themes, in the depiction of events important for the storyteller, and the audience as well as in social values expressed in plots and characters. The same is true of written literature which anthropologists can use as a source in the study of modern literate societies (Spradley and McDonough 1973).

Folklorist Henry Glassie provides a good example of the study of verbal art in a real life setting, and shows how oral performances should be described to reveal the devices employed for effective oral performance. The place in which Glassie worked, and from which this example is taken, is the community of Ballymenone in Northern Ireland. Glassie says that his task was not to write just another ethnography but rather to say something about the workaday world of other people, in this case to describe the *ceili*, the night time gathering of neighbors when the day's work is done and when they meet in someone's home to talk and to tell stories. In fact, the story is the center piece of the *ceili*, and the storytellers are "old men, rarely gifted in storytelling, exceptional but not marginal people."

The stories they tell are "narratives carefully ordered to do the serious work of entertainment, pleasing their listeners in the present, then carrying them into the future with something to think about." The dynamic of storytelling, he says, involves more than dramatic alternations of one mode to another. It also involves repetitive opening words of lines and silence and tones of voice which operate as conspicuous cues to the poetry of the tale. In written poetry you read meter and rhyme. In storytelling the tellers sometimes repeat the first words of a sentence, as well as streamlining their grammar and forming parallel constructions, then hurry the "lines" together without beats or shots of silence. At other times they will shift lines into phrases or even words, separating them into phrases or even words, isolating them for power with alliteration or even rhyme. And when decorating their stories, tellers use key words that are "deep in meaning, repeated and peppered through the discourse to unify the whole."

Those are some of the devices story-tellers use to spin their narrative in an effective manner within the context of an evening gathering of neighbors, in the comfort of a village home, to transport the listeners into other experiences, often old and retold stories, but stories worth telling again. Simply putting the words of the story on paper does not do justice to the storytellers' art, for transcription means translation into another medium with its own conventions of presentation. Nor does translating from the oral to the written medium indicate the kinds of devices used in oral performance that make the narrative effective. Transcription, therefore, should not make the narrative look like written prose, but rather to make it look like sound. That means leaving white spaces on the page to denote silence, as well as some kind of mark, like a diamond sign, to "indicate a smile in the voice, a chuckle in the throat, a laugh in the tale" (1982: 40–41).

A method of recording and presenting oral poetry and oral narrative performances called ethnopoetry has been worked out by Dennis Tedlock and Dell Hymes, in which performances are represented in printed form in such a way (along the lines described above) that assures that elements which constitute a part of the form of the presentation are not lost in transcribing them by using the conventions of writing. Tedlock draws an analogy with music where the recording of musical performance in the form of a score indicates such important features as hesitations, differences in intensity, loudness, tempo, and tone of voice are represented in a system of notation to convey the aesthetic qualities of the performance. Ethnopoetics, therefore, is a convergence point between disciplines since it is a subfield of ethnography and thus sociocultural anthropology, but also of linguistics, philology, literary studies, stylistics, and folkloristics.

The ethnographer studying verbal art also observes the occasions on which various genre are employed, such as free time and for recreation, for rhythmic activities such as hoeing, rowing, grinding, and rocking the baby to sleep, as well as for ceremonial occasions and ritual in the form of chants and hymns. Poetic genres of verbal art are epics, panegyric odes (most highly developed in Polynesia and Africa). Poetic language is most often performed with a musical co-text in the form of songs of various kinds such as ballads, lyrical narratives, laments, and lullabies. There are also different poetic forms performed with music for different groups, such as music for the elite, women's songs and songs for different occupations, as well as songs for different occasions such as festival songs, funeral, war songs, and so forth. Singing is often a component of ritual together with accompanying dances.

MUSIC

Music is everywhere, another of those universals whose origin is probably rooted in the earliest days of the species and which like all other cultural systems varies in form from one culture to another and over time, each evolving in response to changes in the social and cultural context, such as technological innovation, greater occupational specialization, and expanded contexts of use, as well as responses to cultural values and collective myths. Music is thus present in the most basic form of chants and songs around a hunter-gatherer's camp fire and in their rituals, to the most elaborate commercial activities of modern complex society, all a fertile field for anthropological investigation.

The scholarly study of music is called musicology. Its precursors go back to classical times and to *ars musica* which was one of the seven liberal arts of the medieval curriculum. The current form of musicology was formulated in 1885 by Guido Adler (1855–1941) who wrote the first compendium of the field. Adler divided the subject into two parts, systematic musicology and comparative musicology, which later became ethnomusicology. His division forms the basis today for the distinction between music theory and ethnomusicology.

Ethnomusicology studies performances on different instruments as well as music performed vocally and for dances, and the way music is performed in different cultural and social contexts. This approach involves both the humanities and the social sciences since research strategies from both of those fields of inquiry are employed in its study of music and its relationship to performer and social context. In this way ethnomusicology is like linguistic anthropology since both of those disciplines deal with a complex sound system involving underlying structures that can be taken in its own terms with its own form of study and its own specialist terminology as well as studied in terms of its place within a cultural and social setting.

Ethnomusicologist John Blacking says that this field, as it has developed, would be inconceivable without the phonograph (and later the tape recorder) which help translate speculations of the simplicity and complexity of musical forms into facts that can be analyzed with more precision (Blacking 1992). The complexity and quality of music across the different genre, he says, no longer support the strict division of folk, popular, and artistic music in musical terms, although those designations are used for recognizing and studying styles and settings.

In the 1960s Alan Merriam placed ethnomusicology on a firm anthropological basis by defining it as "the study of music in culture" and conversely "the study of music as culture," which includes both Western and non-Western traditions (Merriam 1969). He calls this approach the anthropology of music in which the focus of study is the people who make music and their purpose for doing so, as well as what the music they make and enjoy means to them. The methods used in such studies are those of ethnography; of listening to and recording performances in their natural setting and asking questions about its use and its meaning. In the 1970s Charles Kiel pursued the study of music in that manner, examining popular music as distinct from folk music, and carrying his perspective to Africa where he studied the music of the Tiv of Nigeria (Kiel 1966, 1979, 2002).

Music can be studied in terms of genres, time periods, and social groups of various kinds, but it can also be studied in terms of whole countries. The United States is one amply documented example. One aspect of American popular music is what might be seen as the minstrel tradition, defined as professional performers with a repertoire composed for that purpose, a specialization that goes back thousands of years to the bards of antiquity and the minstrel performers of the Middle Ages, some itinerant and others attached to noble households, who accompanied their songs and recitations on string instruments. This differs from folklore where songs are traditional and transmitted by anyone through the oral medium, yet the line between the two, as in many different areas, is not always clearly demarcated, and as society changed, much of what had been performed in the purely folk medium fed into popular culture. This tradition represents a general realm of activity that would grow in response to changing social, cultural, and economic conditions.

By the nineteenth century a large, literate, and thriving middle class had evolved, creating a market for commercially produced songs in what has become known as popular (as opposed to folk) culture. One of the earliest song writers in the emerging popular music market in the United States was Stephen Foster, who composed his many popular songs in the mid-1900s. His songs were sold in the form of sheet music for middle class families who had pianos in their homes and who played and sang the songs in their parlors for evening entertainment. In the latter half of the nineteenth century minstrel shows were popular. In those performances singers sat in a semi-circle on a stage and sang songs, played musical instruments, and told jokes with the coordination of a central figure known as the interlocutor. The most popular was the black-faced minstrel, but there was also a version of the same type, not in black-face, that was called Irish minstrels. Vaudeville was also a popular form of entertainment from the 1890s to the 1930s. These were theatrical productions that featured songs, dancing, and comic performances together, one after another in a single show. Vaudeville companies traveled from town to town performing on stage in theaters. Vaudeville for its time was considered "the heart of American show business" and was fondly remembered by performances in the 1940s and early 1950s.

Polka was another genre in nineteenth century American music that won a wide following in the United States and that is still enjoyed today, performed in public concerts and for festive occasions. Polka originated in Bohemia in the mid-nineteenth century, spreading across Europe and into the United States, first brought to America by Polish immigrants. Its popularity was widespread but it was most concentrated in the Midwest in the area of heavy German settlement, an area where several styles developed. The polka was also brought to Mexico where it mixed with native styles of music to form the variety known as *el norteño*.

In the late nineteenth century the music publishing industry, which produced songs sold as sheet music, became centered in an area of New York City known as Union Square or Tin Pan Alley. The music produced there was formulaic, and many of the songs were sentimental ballads which were popular throughout the country. Another form of professional music in popular culture was the operetta which began in mid nineteenth century France. A well known composer in that genre was Jacques Offenbach. Richard Straus II was a leading figure in Austria-Hungary where the Viennese style of the operetta developed. In all, Strauss composed sixteen operettas; the best known was *Die Fledermaus*, the most performed piece of that genre, as well as waltzes and polkas and marches which also found a place in the repertoire of American popular music.

The operetta was also developed in England in the 1860s, most famously in the operettas of Gilbert and Sullivan which combined dialogue with songs in telling a story. In the early twentieth century musical stage performances were produced in the United States influenced by Gilbert and Sullivan and by the Viennese style with such composers as Victor Herbert, Sigmund Romberg, and Leonard Bernstein. The operetta gave way to the musical theater, known simply as the musical in the 1940s. Since Broadway was the location of this genre, such performances became known as the Broadway musical. These are stage performances combining dialogue, motion, and songs and which deal with the themes of love and other emotions. One of the earliest and still popular musicals was *Showboat* followed by *Porgy and Bess* and *Oklahoma* and many more.

Musicals played in theaters across the country and later appeared in film versions. Songs from those shows were widely disseminated by phonograph record and radio, many becoming standards in the popular musical repertoire. Among the major composers were George Gershwin, Irving Berlin, and Jerome Kern whose songs comprise what is known as *The Great American Song Book*. The field of popular music has grown and flourished since then, producing such genre as rock in the 1960s and Soul Music, as well as a number of other forms which attract large audiences.

Another important part of the American musical tradition is religious music in the form of gospel music, the Negro spiritual (both with their roots in the black community), and hymns in the European tradition, many imported from Europe, and others composed in the United States, some inspired by revivals. Both spirituals in black churches as well as hymns sung in both black and white congregations are a part of church services and are also sung elsewhere for inspiration, expressing not only a religious message but also reflecting the hopes and aspirations of the people who sing them.

There are four genres of uniquely American popular music with their roots in different ethnic folk traditions. One was from the British Isles. These were folk songs brought by eighteenth century settlers to the British colonies, lingering longer in the Appalachian Mountains than elsewhere, and discovered by folklorists in the late nineteenth century, which soon became a field of folklore scholarship. As radio and vinyl recordings replaced traditional in-home singing, and as song writers exploited the market created by the new music, the old folk tradition developed into a lucrative commercial enterprise called Country Music with its center in Nashville, Tennessee, a music industry with professional singers and bands producing new songs and new sounds, but with themes, melodies and instruments (the guitar, banjo, and mandolin) taken and adapted from the folk tradition from which it had sprung. In the 1960s there was a brief but highly popular folk revival where some of the traditional songs became known to a wider, younger audience.

Another stream of American popular culture with its roots in an ethnic folk tradition is Latin music. The folk and popular base of this genre was the *corrido* and other music of the Mexican border and of Northern Mexico and later with the sounds of the Caribbean and other forms of Latin American music added to it. Latin music is heard today by people mainly of Latin American descent in the United States where its sounds and its leading singers are well known. Also entertainment industry publications like *Record World* and *Billboard* list Latin hits on their charts. Latin music is heard in concert, on the radio, and on television, and outside its primary community, also in night clubs, in recordings, and as music piped into public places.

Another form of popular music with folk roots is the Cajun and the Zydeco music of southern Louisiana with its origins in the French colonial culture of the region. The Creoles of Louisiana are the descendants of French colonists and their slaves, Creole meaning born outside the country of ethnic origin (whether France or Africa) thus including whites, blacks, and mulattoes born in Louisiana and originally speaking some variety of French, and with uniquely French based varieties of culture. Another element of the Louisiana colonial French population are the Cajuns, descendants of refugees from Acadia, now Nova Scotia, who were driven out by the British after the Queen Anne's War of 1701–1714, refugees who eventually joined the French population already in Louisiana. Their music, known as Cajun music, is rooted in ballads from Acadia and influenced, as the genre developed, by American popular music, especially country music. Zydeco is another genre in that tradition, the music of the black Creole population as based on African slave songs of the nineteenth century but with common Acadian roots as those of Cajun music, and influenced by that genre as well as by blues and jazz.

Those folk-based traditions of popular music are well known, but are restricted to certain audiences, the country music audience being by far the largest. Another genre of music with American folk roots, but with quite a different history, is jazz. Jazz emerged in the late nineteenth and the early twentieth centuries in different places in the South but linked by common Afro-American gospel music, spirituals, and blues and with European musical elements. This genre consists of many styles and is thus somewhat difficult to define. Its chief characteristics, however, are performance orientation, improvisation, syncopation, group participation, and the development of an "independent voice." Among its many styles are Dixieland, Ragtime, and Cool Jazz. The latter developed from the end of the 1940s and into the 1950s. It was led by Miles Davis and includes innovative musicians such as Chet Baker and Dave Brubeck. Fusion refers to the uniting, in the late 1960s and 1970s, of jazz and rock. It is characterized by mixed meters, odd time signatures, syncopation, and complex chords and harmonies.

American symphonic music is composed in America but in the European tradition of classical music. It goes back to the eighteenth century, but took on a distinct form in the twentieth century. George Gershwin combined several traditions in his compositions including jazz and African-American music. And Aaron Copland, Charles Ives, and others incorporated American folk music from the Appalachians and from the Great Plains in their work. Composers such as Leroy Anderson and Morton Gould also wrote music in the symphonic jazz style. Leonard Bernstein used mixed non-tonal and jazz in his classical compositions and Charles Ives, John Cage, and Henry Cowell were among those who composed in an experimental vein.

As the cinema developed, so too did the symphonic music that forms a part of cinematographic productions, designed as themes and as background and incidental music to accompany action on the screen. Erich Wolfgang Korngold was one of the founders of film music, and among the best known composers are Max Steiner (*Gone with the Wind*), Alfred Newman (*How the West Was Won* whose overture includes a medley of folk songs), Dimitri Tiomkin (*High Noon* and *The High and Mighty*), John Williams (the *Star Wars* series), and Henry Mancini (*The Pink Panther* composed in the jazz idiom), and *The Molly McGuires* (composed in Irish modal harmony and played with full orchestra and with the penny whistle, the traditional Irish accordion, and the Irish harp).

This brief summary of American music reveals its different ethnic origins as well as part of the varied ethnic makeup of the nation, as expressed in one part of the national culture—in its living musical heritage, a heritage which reveals the different genres and styles which have developed and which reflect class structure, different forms of entertainment, and the transformations wrought by technological innovations such as the phonograph, radio, and television in a progression which has led from middle class parlors and front porch folk singing and festivals and informal gatherings to million dollar industries, all rich topics of investigation for the discipline of the anthropology of music.

CINEMA

The cinema is a modern art form, the technology originating in Europe but developed as a commercial enterprise in the United States. As entertainment and as an art form it has diffused outward to all countries of the world. Like symphonic music and the theater, the making of films requires the co-ordination of several activities under the supervision of a director. Also films—like literature, oral traditions, and graphic and plastic art—both reflect the culture in which they are made, and reinforce its collective images, values, and myths. One well known example is the Western genre which appeared at the beginning and which, for a time, dominated the American film industry. It even spawned an Italian version (the Spaghetti Western) with Western scenery, the Western time frame, traditional Western themes, and featuring American actors, but with Italian composed music and an Italian cinematographic style.

In the early days of the industry the Western was a continuation on film of the nineteenth-century Dime Novel, a genre of mass popular culture. Formulaic in nature and shaping fictional and real characters and events to meet the demands of public entertainment, and in so doing, creating vivid images and helping make legends of historical figures. As the genre grew it became more sophisticated, combining literary, graphic, and musical components not only for entertainment purposes but also as the vehicle of artistic expression. The central theme of the Western was the myth of the frontier, yet the frontier restricted to the last phase of its nearly five centuries long history; a myth which formed a part of the way the nation imagined itself and an affirmation of some of its basic values such as individualism, perseverance, overcoming obstacles, and eventually winning.

The artistry and craftsmanship of Westerns, like films of other genres, became increasingly refined as the industry developed. Eventually the genre diverged, producing along with the standard fare stories, that which depict human situations that could be encountered outside the Western context, but integrating them with the sense of time, place, and flavor which set the genre apart. One example is John Ford's 1948 classic *Fort Apache* which depicts the tension between two figures, the officer who knew the situation on the ground and who respected the Indian adversary, played by John Wayne, and the proper by-the-book commander played by Henry Fonda whose rigidity led to his downfall, the kind of conflict encountered in many other contexts. *New York Times* film critic A. O. Scott shows how that tension still exists in the current military engagements which the United States is carrying out abroad (www.youtube.com/watch?=aus0iQO_a4Y).

Another classic Western of this type was the 1952 film *High Noon* with Gary Cooper and Grace Kelly. The story deals with conflicts between religion and civic duty, courage and civic cowardice, and the very real threat of unmitigated violence in which ideals must sometimes be sacrificed for survival. Another film of that nature is *The Man Who Shot Liberty Valence* (1962) with James Stewart playing an idealistic lawyer newly arrived in a frontier town, and (again) John Wayne as the old hand who knew the territory. Liberty Valence, played by Lee Marvin, is a sadistic outlaw who terrorizes the town, and although the lawyer knows nothing about guns or gun fighting, he meets the villain in a classic showdown in the street. The outlaw fires and wounds the lawyer who shoots back but misses. At the same moment the character played by John Wayne fires from the shadows and kills the outlaw. Later the lawyer is hailed as a hero and is elected to public office, eventually becoming governor when the territory becomes a state, then senator, and eventually an ambassador.

The lawyer knows the truth of the incident but does not disclose it until he and his wife return to the town where the incident occurred, to attend the funeral of the man who really shot Liberty Valence and who had been their friend. Before leaving for Washington, the lawyer tells the true story of the incident to the editor of the local newspaper. When he asks if the editor will print the true story, the editor says, "When you have a choice between the legend and the truth, you print the legend." On the train back to Washington, the conductor informs the couple that the railroad dispatcher will hold the next train at the junction for his convenience. When the lawyer (now an esteemed statesman) thanks him, the conductor says, "Nothing's too good for the man who shot Liberty Valence."

In this film we see courage and honor, standard motifs of the genre, together with comments on the public and the press and the conflict of values which are so much a part of the modern world, presented cinematographically in a Western setting in a masterful way. When the film first appeared, the French journal of the cinema, *Cahiers du Cinema,* featured it in a cover page article titled *L'Americain par Excellence.*

Other commercially produced entertainment films can show in much the same way as do ethnographic films, the way of life of certain segments of society. Among the many examples is the 1955 film *Marty*, from a screenplay by Paddy Chayefsky about an Italian American butcher showing life at that time and on that social level in the Bronx. Another is the 1982 film *Tender Mercy* which takes place in a rural Texas town about simple people at that time and in that place which reveals a way of life in which community, religion, and family are important. *About Schmidt*, 2002, is a comedy-drama loosely based on a novel of the same name by Louis Begley which tells the story of a retired insurance salesman and reveals much about the middle class in the American heartland.

In 1964, a novel by Greek writer Nikos Kazantzakis, written in 1946, was made into the British-Greek co-production *Zorba the Greek*, a film that gives us a look into Greek folk culture as well as traditional Mediterranean gender roles and the importance of a code of honor and a guide to personal behavior which was essential to life in that part of Europe. Another European film, the 2003 German film *Schultze Gets the Blues*, gives us a glimpse of a distinctive American region from the point of view of a foreign visitor. Schultze is a salt miner from Eastern Germany, who after the mine shuts down, does not know what to do with his life. His hobby is playing traditional polka on his accordion. One day he hears a zydeco song on the radio. He quickly turns it off, then thinks about it for a moment, then turns the radio on again and listens to the sounds of that exotic music played on his instrument. When his club sends him to a music festival in the German-settled region of New Braunsfels, Texas, he goes instead to the bayou country of Louisiana where he makes friends and enjoys the folk life and music of the region.

Documentary films also provide information on language and culture as well as do television productions and advertisements of all kinds designed to catch and capture the attention of a particular public. Newspapers, magazines, diaries, text books, and popular and "high" literature, what has been called textual culture (essentially the kind of material Ruth Benedict used for her ethnography from afar), all provide a rich source of data revealing the shared images, myths, and values of a people as well as revealing their aesthetic value and the skill, the craftsmanship, and the artistry of those who make them.

A different approach to Hollywood films, seen from an anthropological perspective, dealt not with the films themselves, their cinematographic features and themes, and their reflection and support of collective images, myths, and values, but with the film industry as it operated in the late 1940s, is Hortense Powdermaker's 1950 *Hollywood, the Dream Factory: An Anthropologist Looks at Movie-Makers*. Her hypothesis is that the social system in which movies are made significantly influences their content and meaning as is the case with folk and popular art, as well as the fine arts. In the case of movie production, movies are conceived and made within the setting of big industry with the co-ordination of many people within a hierarchical structure of management. The methods Powdermaker used were those of ethnography in a study which stood at the beginning of a trend that took anthropological methods and theories into the study of occupations within the wider complex society of the modern world.

FOURTEEN

Linguistic Anthropology

The physical, cultural, and linguistic diversity of the world's population became the focus of what would grow into anthropology. The systematic study of all of those human features was thus included in the scientific holistic discipline as it developed in America. Franz Boas, the founder of the four fields approach, saw the study of language as an integral part of that undertaking when he wrote that the study of language had already reached the scientific stage, predicting that "the scientific understanding of man will in all likelihood grow from our understanding of language" (1939: 105–106).

In 1911, in the introduction to the *Handbook of American Indian Languages,* he laid out what might be considered the programmatic statement of linguistic anthropology in which he summarized phonology and grammar, the essentials of structural linguists, which anthropologists must know in order to describe the many languages of the world in a scientific manner. He also stressed the point that each language had its own internal structure and must be described in its own terms and not by fitting them into a model based on the European languages. He also discussed the genetic relationships which was one way of classifying the languages of the world, another task of the linguistic anthropologist. In that regard he also discussed the way languages in contact influence one another.

He also saw the importance of studying language on the discourse level both for what it can tell us about language and what we can learn about the culture of which it is a part. In this regard he insisted that oral traditions be recorded in phonetic transcription and published and archived in the original language. Such texts would later give us information on native languages while they were still in use for study at a later time when those languages are no longer spoken. He also said that not just oral texts should be studied for their linguistic value, but also for what he called "naïve outpourings" of the natives (he gives the Eskimos as an example), such as newsletters which they wrote for informing people of local events, as well as other writing by natives in their language which contain much valuable linguistic and cultural information.

His view of linguistic anthropology included verbal art in the form of oratory and poetry. In the case of oratory he says that speeches should be recorded and preserved in the original language, and that investigators should examine their method of composition and the devices employed to reach oratorical effect. In the case of poetry he noted the form of rhythm, the adjustment of text to music, the use of language for poetic effect as well as the imagery and the use of metaphors among the numerous elements involved in the investigation of the style of poetic forms.

Boas also said that "from practical, as well as from theoretical points of view, the study of language must be considered as one of the most important branches of ethnological study, because, on the one hand, a thorough insight into ethnology cannot be gained without practical knowledge of language, and on the other hand, the fundamental concepts illustrated by human languages are not distinct in kind from ethnological phenomena; and because, furthermore, the peculiar characteristics of languages are clearly reflected in the views and customs of the peoples of the world" (Boas 1966: 69). For those reasons he said that ethnographers should learn the local language and illicit information in it directly from native informants.

Boas, clearly in the Humboldtian tradition, also says that an extended classification of experience underlies all articulate speech, a fact which he says, "is a fundamental trait of thinking," yet the classifications revealed in linguistic units (words, the roots of words, and affixes) do not conform to the same principles of classification. Instead they vary according to the chief interest of the communities which use them, and "where it is necessary to distinguish a certain phenomenon in many aspects, which in the life of a people play each an entirely independent role, many words may develop, while in other cases modifications of a single term may suffice" (1966: 22). He illustrates this principle for the vocabulary with his well-known example of Eskimo words for snow.

Boas's work in American Indian languages illustrates that perspective which, as Dell Hymes observes, represents "the total scope of American linguistic anthropology until the present time became defined or adumbrated" (1964: 9). Also, Boas's formulation of linguistic study as one subdiscipline of holistic anthropology summarized the converging tendencies already at work in the nineteenth century and which would continue to build throughout the twentieth century in the field of linguistics (chapter 5).

Edward Sapir (1884–1939) was another founder of linguistic anthropology, a theorist, and above all a meticulous field worker who made substantive contributions to both anthropology and autonomous linguistics. In historical linguistics he proposed the theory of language drift and he added a fourth structural type, polysynthetic languages based on features of some North American Indian languages, to the established types of isolative, agglutinative, and inflectional languages. Sapir also examined the affective element in language which he described as the "feeling tones" which are associated with words.

In the last chapter of his book, *Language*, Sapir looks at language and literature, a relationship discussed by Boas (mainly in terms of oral traditions) and continued as a part of contemporary linguistic anthropology. Every language, says Sapir, is itself "a collective art of expression" which contain, concealed within it, "a set of esthetic factors—phonetic, rhythmic, symbolic, morphological—which it does not completely share with other languages" (1921: 225). Writers must use the native aesthetic resources of their languages but, he says, "the language is ready, or can quickly be made ready, to define the artist's individuality" (1921: 231).

Sapir also worked within the broad Humboldtian tradition from which he developed the theory for which he is best known. It can be briefly described in a much quoted passage

> Human beings do not live in the objective world alone, nor alone in the world of social activity as ordinarily understood, but are very much at the mercy of the particular language that has become the medium of expression for their society. It is quite an illusion to imagine that one adjusts to reality essentially without the use of language, and that language is merely an incidental means of solving specific problems of communication or reflection. The fact of the matter is that the "real world" is to a large extent unconsciously built up of the language habits of the group. No two languages are ever sufficiently similar to be considered as representing the same social reality. The worlds in which different societies live are distinct worlds, not merely the same world with different labels attached (1929: 209).

Benjamin Lee Whorf (1897–1941), a student of Sapir's, further developed this idea into what is known as linguistic relativity, also the Sapir-Whorf hypothesis. There are two versions of this hypothesis. One is the strong version that holds that the structure of a language determines thought, and that linguistic categories limit and determine cognitive categories. The weak version holds that linguistic categories do not determine thought but that they do indeed influence thinking as well as certain kinds of non-linguistic behavior. The Sapir-Whorf hypothesis was for a while a major focus of linguistic anthropology which was discussed, tested, and debated. Today linguistic anthropologists generally acknowledge that the structure of a language plays a part in shaping the world view of its speakers and in influencing them in non-trivial ways, but this is no longer a central focus of research.

Another major figure in shaping linguistic anthropology was Alfred Kroeber whose voluminous work in all branches of anthropology (except physical anthropology) has exerted much influence on the discipline (Hymes 1964: 689–707). Kroeber wrote that he had come from humanistic literature to anthropology through the gate of linguistics. He describes Boas's course in linguistics, where students uncovered linguistic structure from texts, as equivalent to the laboratory method which introduced students to an attitude of independent research as well as special training for later field work.

Kroeber's linguistic work ranged from the description of individual North American Indian languages, to genetic classification, structural typology, and lexicostatistics—the statisfied study of vocabularies for historical purposes—as well as aerial relations among different languages. He was also interested in semantics at a time when attention was focused on structure. He credited lexicostatistics with redirecting attention to meaning in language which he described as "one kind of content, one body of phenomena, which language and culture indubitably share" (Hymes 1964: 698).

Kroeber said that language is a part of culture, sometimes a significant part, and that linguistic data are also cultural data with no necessary chasm between them. He also insisted on the cognitive importance of kinship terms, which Hymes describes as "perhaps the language-in-culture problem par excellence in American anthropology." Kinship terms are a prime example of this relationship, for they are part of language with their own degree of autonomy from other parts of culture. Yet as language and as cultural categories kinship terminologies are patterned, each worked out in its own way, and both unconscious systems of classificatory thought. In regard to the Whorfian hypothesis Kroeber felt that it would one day be proved both "true" and "false" on different levels, but only when a new approach is devised to answer what he called "this intriguing and important problem" (Hymes 1964: 698).

Hymes says of Kroeber's linguistic work that it "carries implicit definitions of linguistic anthropology as, simply and broadly, study of language within an anthropological context, and linguistic anthropologist as one who uses linguistic techniques and data to answer anthropological questions" (1964: 703). Almost any general anthropological question can be asked of language, and some questions of culture cannot be answered without the aid of language. This approach to language, together with the other aspects of linguistics, was introduced by Boas and enlarged and passed on by Sapir and Kroeber to form the basis of the current field of linguistic anthropology as a part of anthropology as a whole.

Interest in the social aspect of language is partly due to the influence of Bronislaw Malinowski who called for the development of an ethnolinguistic theory in connection with ethnography. In 1923, he addressed that issue in an article titled "The Problem of Meaning in Primitive Languages." And in 1935, he expanded his views on language in his *Coral Gardens and Their Magic*, Part II. The rise of sociolinguistics and the influence of ordinary language philosophers (chapter 6) also affected linguistic anthropology as seen in the work of Dell Hymes and the interactional sociolinguistics of John Gumperz (1982a, 1982b). In 1962 Hymes wrote an essay titled "The Ethnography of Speaking" which is concerned "with the situations and uses, the patterns and functions of speaking as an activity in its own right" (1968: 101).

This approach drew on previous work by Malinowski, the Prague School, and work to that point in the cognitive approach to culture. In that regard Hymes presents a redefinition of linguistic competence—what one must know about the structure of a language to generate sentences—as communicative competence which includes knowledge of the linguistic code, but also what one must know about the conventions of speaking in order to use it in the appropriate situation and manner. In 1964 he refined this approach in a book titled *Toward Ethnographies of Communication* and in 1972 he and John Gumperz co-edited a book which included their own articles on the social aspect of linguistic anthropologists along with articles by others on topics relevant to that approach, titled *Directions in Sociolinguistics: The Ethnography of Communication*.

The ethnography of communication looks at the total act of linguistic communication. This includes the setting of speech transactions and the participants (speakers, addressees, audiences) as well as speech events and speech acts, the tone and manner of the transaction (serious, joking, ironic, etc.), the purpose of the event and the intention of speakers in uttering speech acts, along with the style, register, dialect (or other language) selected from the speaker's repertoire when speaking, the genre employed interview, conversation, and gossip as well as the norms of interaction. With this approach we see not only a merger of sociocultural anthropology and linguistics but a clear formulation of the convergence trends discussed in chapter 6.

In sum, language can be studied on its own terms (grammar, phonology, and lexicon) apart from its functions, other than those of encoding and transmitting messages. Linguistic anthropology, beginning at that point, further views language as a mode of thinking, a cultural resource, and a powerful social tool, and speaking as a patterned cultural practice. The emphasis on linguistic anthropology, says Alessandro Duranti, is "on communicative practices as constitutive of the culture of everyday life" (Duranti 1972: xv, 1–2) and in this way it is pursued cross-culturally, revealing in its conception, its history, and its practice the interdisciplinarity that is characteristic of anthropology as a holistic science.

FIFTEEN

Culture Areas

The Case of the Central Andes

One of the early interests in anthropology was the study of culture areas (such as the North American Great Plains, the Subarctic Area, and in Germany the *Kulturkreis* School). Culture Areas are geographic zones characterized by environmental homogeneity and by a number of shared cultural and social traits, despite differences in ethnic identification and language. Another interest among anthropologists is the study of larger regions such as Latin America and Eastern Europe. The American Anthropological Association has special sections for such regional interests.

The Central Andes is a notable example of both a culture area and a modern region. First, it is one of the "cradles of civilization," with a long culture history retrievable by means of archaeology, with a five hundred year long written record, and with a large traditional agrarian sector which retains many cultural traits from earlier times. It is also a part of the larger region of Latin America. And within that region it shares traits with Mexico and Guatemala, for both were homes of the only civilizations to emerge in the Americas, both were conquered by the Spanish in the sixteenth century, and both have similar post conquest histories, and today large indigenous and mixed blood (mestizo) populations. Those similarities have led some to call Mesoamerica and the Central Andes "Indoamerica."

The Andean countries are also a developing region which means that they are undergoing steady change in the direction of closer integration in a globalizing world. They are also one of three high mountain regions in the world, with similar environments to which native populations have both physically and culturally adapted. The Central Andes, therefore, is a place where cultural and regional studies of different kinds meet, and where all four fields of anthropology come together and interact with neighboring disciplines.

THE HIGH MOUNTAIN ENVIRONMENT

The Andes is the longest continental mountain chain in the world, running like a backbone from north to south down the western side of South America, from the Caribbean to Tierra del Fuego, a distance of some 4,300 miles. The highlands are some 120 miles wide throughout most of their length, but expand to some 400 miles in Bolivia (between 8 and 20 degrees south latitude). The average elevation is around 4,000 meters above sea level (13,000 feet), with the highest peaks rising to above 6,000 meters (20,000 feet). The highest is Aconcagua in Chile at 6,962 meters (22,029 feet) which also makes it the highest mountain outside Asia. (The highest in the world is Mount Everest in the Himalayas at 8,848 meters (29,029 feet).

The Andes form three distinct regions, the Northern Andes of Colombia and Venezuela, the Southern Andes of most of Chile and Argentina, and the Central Andes of Ecuador, Peru, Bolivia, and the northern part of Chile and Argentina. The Northern Andes are narrower and with less elevation than in the central portion. Also its vegetation, which varies with elevation, is typical of moist tropical climates. The Southern Andes, which extend to the cooler southern latitudes, is more typical of mountains in temperate zones. The Central Andes extend from just north of the equator to the Tropic of Capricorn, which places them in the tropics. Due to their elevation and different patterns of humidity, the vegetation of the region ranges from warm and arid to subarctic to tropical rain forest vegetation.

There are fifteen mountains in the Central Andes that rise to over 6,000 meters above sea level. Among them is Chimborazo in Ecuador, at 6,962 meters (20,577). This mountain is considered the farthest mountain peak from the center of the earth, for the earth is not completely spherical and bulges a few miles close to the equator. The highland region is separated into north-south running mountain ranges, the Maritime Cordillera in the west, and the Eastern Cordilleras in the east. The two mountain chains are separated by plateaus and intermediate depressions. Several high plateaus characterize the southern portion of the region. The most extensive is the broad Altiplano of southern Peru and northern Bolivia, with an average altitude of 3,750 meters above sea level, making it the second highest in the world after the Tibetan Plateau in the Himalayan highlands. Another distinctive feature of the Altiplano is Lake Titicaca which lies at 3,812 meters on the border between Peru and Bolivia. The lake is 120 miles long and covers an area of 3,200 square miles, and is the world's highest navigable lake.

Besides the variation of altitude from sea level to over 6,000 meters, the region is also characterized by two intersecting moisture gradients ranging from extremely humid to extremely arid. The North-South gradient begins in the northern part of the Central Andes, becoming drier as one travels south and ending in the Atacama Desert in the northern Chilean highlands and on the southern Peruvian and northern Chilean coast. The east-west gradient goes from the arid western slopes of the Maritime Cordilleras and the coast, to the eastern slopes of the Eastern Cordilleras which fall to the tropical Amazon lowlands. Those two intersecting gradients, together with a temperature gradient, a product of elevation, have made the Central Andes ecologically quite diverse.

In the west, a narrow coastal strip runs between the Pacific and the Maritime Cordilleras which widens out into the Nazca Plain in southern Peru. Beyond the shoreline the surface of the earth plunges even farther into the Robertson Trench, a formation that lies as far below sea level as do the highest peaks above the ocean's surface. Also the prevailing rain-bearing winds blow not across the continent from the west, but rather parallel to the coast, together with the south-north flowing Humboldt Current. As a consequence, the coast is arid from southern Ecuador to near Concepción, Chile.

The upwelling of cold water in the Humboldt Current forms a cloud cover over the ocean and the coastal strip, and seasonal fogs moisten the arid coastal hills in central and northern Peru which blossom once a year in the form of fog meadows called *lomas*. Much drier is the Atacama Desert which lies along a 600 mile strip of coast and reaches to the Altiplano with an average elevation of 4,000 meters above sea level. The Atacama is the driest non-polar desert, and perhaps also the oldest desert, in the world. Its bleak landscape has been compared to that of Mars.

The eastern slopes, known as the *montaña*, fall to the Amazon Basin, a vast lowland area as large as the United States west of the Mississippi and covered by the largest rain forest in the world, a stark contrast to the arid opposite side of the mountains which includes the world's driest desert. The prevailing rain-bearing winds on the eastern side of the mountains blow from east to west bringing a high volume of rain and forming a band of cloud forest along the forested east-facing slopes, and topping the peaks of the Eastern Cordillera with permanent snow. The snow line varies from the moister north to the less moist south, and is canted, with a lower snow line on the eastern than on western slopes. Snow line elevations range from 4,500 to 4,800 meters in Ecuador and northern Peru and 4,800 to 5,200 meters in southern Peru and Bolivia. Seasonal rainfall diminishes from east to west, stopping at the Maritime

Cordillera and thus never reaching the already arid western slopes and the coast.

The original plant cover of the central Andes is a mark of its ecological diversity. In 1947, geographer L. R. Holdridge devised a classification of the world's plant formations, and in 1958, another geographer, Joseph A. Tosi, refined Holdridge's classification by including the climatic basis for the variation such as temperature, precipitation, and estimated potential evapotranspiration. He then applied the system to Peru, relying on the description of plant life in Peru published in 1911 by Augustin Weberbauer (*Die Pflanzenwelt der PeruanischenAnden*).

Of the thirty-seven life zones identified by Holdridge, all but three are found in the Peruvian Andes. The key factor here is elevation. The slopes and valleys of the Andes are characterized by vertical life zones, one next to the other, providing an extremely varied ecology in compressed spaces. In this respect, geographer Isaiah Bowman has observed that in places on the eastern side of the Andes of Southern Peru, it is possible to pass through the climates of the world in a single day (Bowman 1916).

At the top of the series are glacial peaks and snow fields and below that in places is a frost desert. Between those zones, and the upper limits of continuous forest, lies a zone known in Southern Peru and Bolivia as *puna* and in Northern Peru and Ecuador as the *paramo* which in Northern Peru is also called *jalca*. The difference between the *puna* and the *paramo* is a product of the decrease in rainfall as one moves south, from plentiful moisture of the tropical north, to the aridity of the Atacama Desert. Temperatures vary between -2 degrees Centigrade at night to from 10 to 20 degrees by day. Three of Holdridge's life zones are found there, from scrub brush in the moist *paramo* to a grass landscape on the *puna* with sedge and rushes in places of poor drainage. The relief is moderate compared to the steep escarpments on the eastern and western slopes and in the intermontane valleys. Most of this high altitude zone is in the form of a rolling plane. Its widest extension is in the wide plateaus of Southern Peru and Bolivia.

PHYSICAL ADAPTATION TO A HIGH ALTITUDE ENVIRONMENT

Anyone traveling to a higher altitude has noticed its effects. At some three thousand meters (11,000 feet) those effects become more severe in the form of fatigue, dizziness, fainting, insomnia, malaise, loss of appetite, nausea, vomiting, body pain, and so forth. Those discomforts are known as mountain sickness or saroche, and their cause is hypoxia, the effects on the body of decreased amounts of oxygen in the air. People who have been born and brought up there, however, do not suffer from those effects.

Newcomers eventually become acclimatized, although they feel the stress with increased activity, but the indigenous population, which has been living above 3,000 meters in the Andes for some ten thousand years, has adjusted to those conditions over time. Acclimatization is the adjustment by an organism to new conditions. Adaptation on the other hand involves heritable genetic change which is part of the evolutionary process. In this regard, studies of the indigenous population have shown massive modifications of the oxygen transport system of the blood, especially molecular changes in the hemoglobin, a protein for carrying oxygen in the body which compensates for the low oxygen content of the atmosphere at such elevations. There are also associated traits such as high birth weights, increased lung volume, increased breathing, and higher resting metabolism. Physical anthropologist Paul T. Baker spent several years studying the effects of living at high altitudes among the Quechua and Aymara-speaking population of Southern Peru. He began his study in the Cuzco area in 1962. In 1964, he extended his work to the district of Nuñoa on the altiplano of Peru, continuing his research throughout the 1960s.

For forty years another physical anthropologist, Cynthia Beall, has studied high altitude populations not only in Peru, but in Ethiopia and Tibet as well. According to her research physical conditions such as resting ventilation, hypoxic ventilator response, oxygen saturation, and levels of hemoglobin among the Aymara of the Andes, and the Amharans of Southwest Ethiopia were significantly different from those of the Tibetans. This indicates that different people have adapted to high altitude conditions in different ways, an example of convergent evolution.

A number of characteristics contributing to a high altitude Andean phenotype have been identified or postulated. (A phenotype is the composite of the observable traits of an individual such as morphology, development, biochemical, or physiological properties.) However, still unresolved is the relative contribution to that phenotype of development adaptation within individuals, and the genetic adaptation within populations in which the individual lives. In other words, are those traits dictated by genetic heritage and if so, in what proportion, or do the traits of the phenotype develop in response to the environment in which individuals are born and raised? One thing is certain. On the basis of studies by physical anthropologists and other biologists, part of the native Andean population has physically adjusted to life in the high altitude zones of the Central Andes, giving us a clearer picture of the population that has inhabited the region for millennia. Cultural adaptation is also a factor in the way Andean civilization has developed, a pattern which can be read in the archaeological record of this distinctive culture area.

ANDEAN ARCHAEOLOGY

The Inca past has always been well known from the historical record, but beyond that there was only myth. It was not until the middle and the end of the nineteenth century that the vast and rich archaeological record was explored. In 1864, Ephraim George Squire excavated the extensive ruins of Pachacamac on the coast south of Lima. Among the late nineteenth century pioneers of Andean archaeology were Thomas J. Hutchinson, Wilhelm Reiss, Alphons Stübel, Charles Wiener, Ernst Middendorf, and Adolph Bandelier.

The interest of those investigators was in reconstructing ancient culture, not in their individual histories as read by means of scientific methods. The era of scientific archaeology began with the work of Max Uhle who between 1899 and 1909 carried out extensive excavations of graves on the Peruvian coast. Uhle excavated Pachacamac on the coast south of Lima in 1896 reporting his findings in 1903. He was the first in the Americas to use a true archaeological stratigraphy in his work, and on this basis he proposed a scheme of three major periods in Andean culture history: Pre-Tiahuanaco, Tiahuanaco, and Post-Tiahuanaco. Uhle's scheme was more firmly established in the mid-1920s by the research of Alfred Kroeber, William Duncan Strong, and Ann H. Gayton. Their work provided the basis for the present chronology of Andean culture history, which has since been refined and extended back into time, as more has been learned about the Andean past.

Field investigation continued in the 1920s and 1930s conducted by Alfred Kroeber, Wendel Bennett, and Julio C. Tello. Perhaps the most important result of that work was Tello's formulation of what he called the Chavin culture, based on stylistic uniformities that appeared over a wide area. As it turned out, the uniformity found in the archaeological record did not represent a single culture, but rather the spread of an artistic and cult-based style that connected different cultures.

Archaeological work intensified in the 1940s when the Institute for Andean Research sponsored various excavations in Peru. One of them was the Virú Project conducted on the Peruvian coast in 1946, the first time that an entire valley, rather than single sites, was systematically surveyed with the intent of understanding its full culture history. In 1953 Gordon Willey published the findings of that study, in which he traced the history of human settlement in the valley, relating settlements to their physical environment as well as to their agricultural needs and to the development of irrigation, population growth, warfare, and the relationship of those settlements with communities in neighboring valleys. The holistic approach which Willey established for that one coastal valley came to characterize the archaeological approach to the Central Andes in general, and provided insights into the nature and the causes of growth of ancient Andean civilization. Since 1955 research in Andean history has greatly expanded.

Continuities which characterize the contemporary rural landscape in the Andes make ethnoarchaeology a useful tool in the study of Andean prehistory. Archaeologist George Miller in 1979 was the first to apply this method in his study of native llama and alpaca herding in the highlands of Southern Peru. He focused on the problem of identifying domestic camelids and their economic uses by analyzing the morphological features of wild and domestic camelid bones, a study that would be useful in differentiating archaeologically the wild from the domestic species. Another of his findings was what he calls "the charki effect." *Charki* (the Quechua word for "dry") is freeze-dried meat prepared at high altitudes by leaving meat out at night where it freezes, then letting it thaw in the sun during the day, producing a nutrition rich and easily preserved meat product. The English word *jerky* comes from the Quechua word "*charki*."

Miller noticed that pastoralists in high altitude herding zones often retained low-meat yielding feet and vertebrae for freeze drying for their own purposes, and used choice portions of meat for trading with agriculturalists. The result is a distinct distribution of meat parts which help archeologists track the movement of meat in trade (Miller 1979). Miller's work was followed and refined by Johnathan Kent (1982) which Lawrence Kuznar and others have corroborated, refined, and expanded (1995, 2001).

ANDEAN CULTURE HISTORY

In order to understand the culture history of an area a chronology must be worked out on the basis of archaeological evidence. Below is a chronology that covers the period from the beginning of human habitation to the Spanish Conquest in 1534. It consists of two stages, Perceramic and Ceramic with each stage divided into successive periods (Lanning 1967: 22–27).

The first humans entered the region as hunters and gatherers via the Isthmus of Panama around 15,000 BC. They spread as migratory food gatherers on the Central Peruvian coast and into the highlands as deer and guanaco hunters moving ever higher as the glaciers melted. One site was found in a cave at an elevation of 4,572 meters (15,000 feet).

Basic changes took place in Preceramic Periods V and especially in Period VI. Cotton and squash were domesticated on the coast giving rise to agriculture and the production of crops and textiles. Potatoes and other tubers as well as llamas and alpacas were domesticated in the highlands of southern Peru and Bolivia, and maize was diffused from Mesoamerica through the highlands and along the coast. The Ceramic Stage

Table 15.1. Andean Culture History

STAGE	PERIOD	DATE
CERAMIC	Late Horizon	1476–1534 AD
	Late Intermediate	1000–1476 AD
	Middle Horizon	600–1000 BC
	Early Intermediate	200 BC–600 AD
	Early Horizon	900–200 BC
	Initial Period	1500–900 BC
PRECERAMIC	Period VI	2500–1500 BC
	Period V	4200–2500 BC
	Period IV	6000–4200 BC
	Period III	8000–6000 BC
	Period II	9500–8000 BC
	Period	c.15,000–9500 BC

begins with the appearance of pottery in the archaeological record, beginning at different times in different regions.

An innovation in the Initial Period of the Ceramic Stage was the appearance of the heddle loom. Also nomadic patterns associated with hunting and gathering were replaced with village life. At the same time there was an increase in public construction. Regional isolation which had characterized the Preceramic Stage was intensified as revealed in the stylistic variation in pottery, architecture, tool types, and settlement patterns. The one exception to increased regionalism was the obsidian trade in the south which had no connection with the ceremonial centers of the region. In sum, the spread of the use of pottery, the heddle loom, cultivation of crops, the herding of livestock, and the spread of village life gave the period a more or less uniform character despite the regional variation.

Around 900 BC the first of the "great styles" appeared as seen in the spread of artistic motifs and of architecture associated with the spread of a cult. This style is known as Chavin after Chavin de Huantar in northern Peru where a major temple associated with the cult is found. The Chavin style influenced the Ecuadorian highlands but was located in northern Peru never reaching the southern Peruvian coast or highlands. In sum, the Initial Period contained and intensified the new ways of life that appeared in Preceramic VI, with the Chavin cult representing their culmination, and at the same time indicating the first trans-regional impulse that would come to characterize the rest of Andean culture history.

The Early Intermediate Period beginning around 200 BC was another time of major change. Valley-wide irrigation developed on the coast, population growth and density increased, and the first large cities appeared, some serving as the capitals of rather large states. This was also a time of warfare as revealed in the excavation of graves and as represented on pottery and textiles which depict battle scenes, bound prisoners and trophy heads. There were two areas at that time where art was particularly well developed and widely used. They are Nazca on the southern, and Moche on the northern coast of Peru. Moche artists were prolific, depicting realistic figures in clay and painted pottery everything about their daily lives and culture, leaving us a picture of the world in which they lived and the world of the supernatural in which they believed. Among their many themes were plants and animals depicting at least thirty-five species of birds, mammals, and sixteen species of fish as well as everyday pursuits in all social classes and occupations. Moche artists also depicted different acts of sexual intercourse, earning for their ceramic the name "pornographic pots." Nazca artists were more interested in gods and mythological beings, but they also showed people in their various occupations.

The Middle Horizon was characterized by the rise of two great empires, the Huari Empire, which at its fullest extension went from Cajamarca in the northern Peruvian highlands to Sicuani south of Cuzco, and Tiahuanaco which occupied the entire Titicaca Basin and the highlands of northern Chile. Goods were exchanged between regions more than before which marked a new phase in Andean culture history. The Huari Empire lasted only a century or two after which it disappeared. Its capital Huari and all other cities in the southern highlands were abandoned and the population scattered over the countryside to live in small towns and villages. Tiahuanaco was also short-lived with the same consequences. The theory is that the Huari and the Tiahuanaco Empires imposed a higher level of authority over numerous polities, a system that worked for a while, but then collapsed, perhaps because they failed to come up with a workable method of tying together diverse regions into a wider multiethnic state. And when the empires collapsed they left behind turmoil and increased regionalization.

The ancient civilizations of Egypt, Mesopotamia, and China left written records giving us some idea of the rise and fall of empires in those parts of the world. Ancient Andean cultures however had no writing system, and there is nothing in the archaeological record that would tell us of the causes of the rise and collapse of the Middle Horizon empires. However Andean oral history, as reported by the sixteenth and early seventeenth century chronicler Guaman Poma (1615) gives a native account of subsequent conditions, which archaeologists call the Late Intermediate Period. "The good settlements were depopulated in fear of war and uprisings," he writes. "They went to settle on the heights and mountains and ravines so as to defend themselves, and they began to build fortresses . . . They built walls and enclosures and houses inside . . .

hideouts . . . and many died." This description is confirmed by archaeological investigations.

That was a time of cultural regionalization, but it was also a time of high technology and the rise of small regional polities which were often at war with one another, but not for territory, since those struggles did not result in the consolidation of military victories into larger political entities.

ETHNIC POLITIES AND NASCIENT STATES

As we saw in chapter 12 that all societies must attend to internal order and to relations with outside groups. The way they deal with those tasks constitutes their political organization. On the level of small equalitarian hunter-gatherer and horticultural societies, social organization is built on highly personalized relations among all the members of the group. Each individual learns his and her position in society as they grow up in it, and the goals of the society are accepted on the basis of tacit agreement. Internal order is therefore maintained through custom and example and when necessary by community retribution. Leaders are those who are recognized as the most capable individuals for deciding how to reach common goals. In some cases there are chiefs for dealing with internal problems and others for organizing and leading warriors, either in defense or for other reasons. Interaction with the spirit world is through the shaman, an individual who feels that he has a special relationship with the other world and not through a formal priesthood.

Larger societies living in agrarian villages, however, have a more pronounced political system where hereditary lords stand at the top of a more stratified social order. Tribute is paid to them due to their higher status and their role in administering group affairs. The lord then redistributes the tribute in various ways to maintain social solidarity. The authority of the lord is seen as his right to command based on his status and leadership functions and on custom. States appear when an agrarian population grows and the social and the economic systems become more complex. The requirements of internal control and external relations become more formalized, crystallizing into a state with an administrative bureaucracy.

The process of developing into a state is pretty well known from the archaeological and historical records of ancient Egypt, Mesopotamia, the Indus Valley, and China, although the causes of the shift from ranked polities to the status of states cannot be exactly specified. The shift however takes place when the top of the hierarchy becomes a government which relies on taxes and agents to support the public goals of the impersonal hierarchy and not on traditional patterns of redistributive and reciprocity. Under such conditions the agrarian rural population becomes a peasantry and the state institutions spread and replace other social and economic forms. Another characteristic of early states is that at some point one state will expand its power and administration over others in the area, thus forming empires which rise and eventually fall in cycles of empire.

What the archaeological record reveals of the Middle Horizon in the Central Andes is the rise of the first Andean states and the first cycle of empire, and in the Late Intermediate Period we see a time of flourishing polities presided over by local ethnic lords and characterized by patterns of reciprocity and redistribution which, as the system ripened, created the preconditions for what would be the last round of native empire in the Central Andean culture area, the rise of the Incan Empire.

THE VERTICALITY OF THE ANDEAN WORLD

In the Late Intermediate Period, the Andean resource base consisted of tubers, of which potatoes are the best known, maize, cotton, alpaca, and llama wool and meat. Coca was also used by placing leafs in the mouth. It is believed that one becomes less tired and feels hunger less when coca is used in that way. Coca was also essential in Andean rituals. Coca originated in the *yunka* or warm valleys of the eastern slopes of the Andes. Maize diffused into the Andes from Mesoamerica and was cultivated on the coast and on the protected slopes of highland valleys up to 3,500 meters. Its cultivation was closely correlated with irrigation agriculture. Also terraces extended its cultivation into the highland valleys to its upper limits. Maize is a warm weather crop. Irrigation made crop rotation and fallowing unnecessary for growing maize. The irrigated coast was the major producer of maize, and the rich guano deposits on the coast were used as fertilizer. Maize was a food, but it was also used to make corn beer called *chicha*. Maize was easy to store for later use.

Maize was also the center of religious rites on the folk level with rituals performed at harvest time, consisting of singing and feasting. During the celebration shrines to Mama Sara, "corn mother," were kept in homes. They consisted of the best ears of corn wrapped in the family's best blanket. Maize was also used in rites of passage; a boy's first haircut and weddings and funerals. It also played a role in local rituals even in zones where it did not grow.

Tubers were domesticated in the high altitude zones of southern Peru and Bolivia and are adapted to the cold climate. Unlike maize, tubers are cultivated without irrigation and on fallowed fields. Food preservation in the cold high altitude zones was by means of freeze drying. One variety of high altitude potato is left outside to freeze at night and thaw during the day to produce what is known as *chuñu* a product that can be stored.

The same thing is done with meat, which is called *ch'arki*, a product that could be stored and transported, as described above.

Rituals associated with the protection of potatoes were also observed in zones where it was grown, but overall maize enjoyed a special status among the crops. The chronicles communicate the feeling that maize was a desirable, special, and even a holiday food as compared with potatoes (1960: 397). Even though tubers had a lower status, they were important as a staple. The number of varieties and hybrids indicate that they had been in use for a long time. And the high altitude based Tiahuanaco expansion shows that tubers were not marginal to state growth.

Coca had been used in the Andes for some eight thousand years as seen in excavations where it has been found in association with mummies, along with spatulas employed in its use, bags for carrying it, gold statues, and depictions on Mochica ceramics. Cotton and the wool of llamas and from the more valuable alpaca were used to produce textiles which not only met the need for clothing, especially in the cold higher elevations, but it also had high decorative and artistic value and was used for social and political purposes. John Murra (1962) says that gifts of textiles from the Inca to the *kurakas* (local ethnic lords) were a way of binding their relationship. And a whole corps of weavers, the *aqllakuna*, was employed in the production of textiles on the state level. Llamas were the main means of transport in the Andes and thus were of high value. Also camelid meat was used for food.

Based on the reading of the Spanish chronicles and post-conquest Spanish administrative documents, Murra (1972) developed a model (what Max Weber called an ideal type) of the social, political, and economic nature of the polities of that period. First was the *ayllu* or the folk community, based on ties of kinship and reciprocal labor arrangements. The relationship between the community and the hereditary local *kuraka* was one which involved a system of redistribution where part of the resources from crops and livestock was given to the lord who then distributed it within the polity. This pattern of redistribution has appeared elsewhere, but in the Andes it took on a different form, for it represented a specific adaptation to the Andean vertical environment.

An ethnic polity had a nucleus in one of the vertically ordered resource zones, which was the most populous area of the polity's territory, the seat of power and the residence of the *kuraka*. If its center was in the highest life zones the group would have ample tubers, quinoa (a native cereal), and llamas (and in the highest zones alpacas), and thus meat, wool, and a means of transportation. In order to gain access to maize, colonies were established in the valleys where maize was cultivated under milder *qheshwa* conditions. For other products they established colonies in the still warmer *yunkas* which in the west lay on the arid land and in the east, the moist banks of the Andes. They raised maize, coca, and hot pepper in both of those zones, and extracted timber from the eastern jungle and fish and guano, for fertilizer, from the coast.

If the nucleus was in the temperate *qheshwa* zone, colonies would be established in the highest life zones, the *puna*, and in the *yunkas* on both sides of the Andes as far as possible. In this way ethnic groups sought maximum use of all the resource zones of the vertical series. The pattern that evolved has the form of an archipelago up and down the mountains. Space between the outlying settlements was not under the control of the ethnic group, and colonies from different groups lived side by side with other ethnic polities and their outliers. There was little or no trade between the groups for there is no trace of market centers. The movement of goods took place within the polity by llama caravan between the center and the outliers within the polity based on ties of kinship, reciprocity, and redistribution.

In managing this complementarity ethnic lords eventually acquired governing and military skills, which were probably quite early translated into strategies for dealing with competing polities. Other groups shared peripheral zones in sometimes tense temporary truces, thus the development of diplomatic skills. And military skills were developed in disputes with neighboring groups and for dealing with marauders who preyed on the outlying colonies. From those conditions institutions emerged, and leaders arose who would know how to administer a multiethnic and multilingual territory.

The idea of empire, once in place, seems quite resilient as seen in ancient China, Egypt, and the Middle East with their pattern of cyclic empires. It was also evident in the West where the empire of Alexander the Great was followed by that of Rome whose ideal persisted throughout the fragmented Middle Ages in the form of the Holy Roman Empire (which has been described as neither holy nor Roman nor an empire) and into the modern period with the title emperor and the trappings of imperial Rome assumed by Napoleon, and in the Roman image that persisted in the colonial empires of eighteenth, nineteenth, and twentieth century Europe.

The survival of the ideal of empire was also apparently the case in the Andes, for as regional states evolved during the Late Intermediate Period, eventually so too did the urge on the part of some of them toward wider political unity. This was seen in the expansion of Chimú in the Moche Valley, a state which conquered the north coast of Peru nearly to what is now the Ecuadorian border. Chimú's expansion was halted when it was conquered in turn by a more powerful state from the southern highlands, the state ruled by the Incas in Cuzco.

The Inca Period marks the final phase of native Andean civilization, the cumulative effects of a long process of sociocultural elaboration. It is that process which Julian Steward examines, bringing to the study of the Central Andean culture area a nomothetic perspective, looking at regularities in the stages of sociocultural elaboration in the emergence of all ancient civilizations. Steward notes the overshadowing importance of ceramics in the archaeological record when establishing chronologies of culture history, but archaeological material also contains much of social, economic, and cultural interest which provides a basis for a developmental chronology. On that basis and within that theoretical perspective, he views Andean culture history in terms of eras that show a steady line of social and cultural elaboration:

The Era of Hunters, Gatherers, and Fishers
The Era of Incipient Farming: Folk Communities
The Formative Era: Emergence of Theocratic States
Regional Florescent States: Priests, Warriors, and Artisans
Cyclical Conquests: Prehistoric Empires

The final stage in this process was the Inca conquest of the entire Central Andean culture area, uniting the different regions and ethnic polities into a single Pan-Andean system, elaborating the institutions and practices that had evolved there in the course of the development of Andean civilization, and extending them to dimensions until then unknown. The Incas were in the process of transforming deeply rooted Andean social, economic, and political ways of life into an unprecedented unity with new structures, until the process was cut short by the Spanish conquest in 1532.

THE HISTORICAL RECORD

The rise and expansion of the Inca Empire, like the two empires before it, left its trace in the archaeological record as seen in the spread of a single style and with constructions built to facility movement and the administrative needs of the empire. Unfortunately we have no Inca documents since Andean civilization never developed a system of writing, one of the features which separate it from other high cultures of the world. However a complex empire that extended over 2,500 miles and contained millions of people requires for administrative and military purposes some way of keeping track of its resources and its population.

What they came up with was not writing, but rather an accounting and mnemonic devise consisting of a system of knotted cords called *quipu,* which they used effectively for census taking, taxation, land use, and assembling and storing calendrical and military information. Messengers also consulted *quipu* to aid their memory when transmitting the oral messages they were delivering. Specialists in the use of the *quipu* formed a kind of clerisy, and the nobility learned to read *quipu* in formal schools known as *yachawasi,* literally "house of knowledge." The Inca equivalents of historians consulted *quipu* when recounting Inca history to the Spaniards. No one knows, however, whether those knotted cords were only numerical or whether they in some way contained narratives.

Our knowledge of the Inca Empire comes from those oral histories as recorded in the second half of the sixteenth century by Spanish chroniclers, as well as from a few brief firsthand descriptions of some ceremonies written by Spanish soldiers who were with Pizarro. The veracity of Indian witnesses is difficult to ascertain at this distance in time, complicated by the natural variation one finds in oral traditions. Also translation difficulties arise as well as the practices of sixteenth- and seventeenth-century chroniclers of often liberally copying from one another without attribution, so that much writing of the time which are considered original "sources" and "documents" are only restatements at third or fourth hand of original testimony, accounts which are also flawed by carelessness and political or personal prejudice. For those reasons John Rowe warns that for an accurate gleaning of the Inca past from the historical record one must use it with discretion (Rowe 1946: 192–197).

Among the most valuable sources of that time was the work of Pedro de Cieza de Leon and Juan Polo de Ondegardo. Ciezo de Leon was an author and a soldier who wrote in the 1550s. He has been described by Rowe as "honest, conscientious and thorough." Polo de Ondegardo was a government official in Cuzco in the 1550s who wrote his reports between 1561 and 1571. With the exception of those two writers the sixteenth- and early seventeenth-century Spanish chroniclers were interested only in the idealized statements of the bureaucratic claims of the Inca state and civilization.

Two native writers, who also wrote chronicles based on oral traditions, give us information not only about the Inca past, but also about everyday life of their time in which they lived, which was the time just after the conquest and thus a continuation of life during the Inca Period. One of those chroniclers was Felipe Guaman Pomo de Ayala (1535–1616), a man from the *kuraka* class who lived near Ayacucho in the Southern Peruvian Highlands. He acted as translator for the Spanish in writing their accounts of Inca history. He also wrote his own account of Inca history and a description of the period in which he lived. Rowe describes it as "a monumental work of 1,000 pages" which contains full-page drawings in ink of daily life for which is perhaps the best part of the text.

The other native chronicler was his contemporary, Garcilaso de la Vega (1539–1616), who was born in Cuzco, the son of an Inca noblewoman and a Spanish conqueror. His history of the Incas was written late in life and in Spain. It became a part of Spanish literature and one of the best known of the chronicles although not the most accurate. However, the best overall history, taken from some of those sources, was written by the Spanish Jesuit Bernabé Cobo around 1653.

The historical record is greatly augmented by Spanish documents of various kinds such as tax records and surveys and other information gathered for the Spanish king by order of Francisco de Toledo who was viceroy from 1569 to 1582. These documents are some of our most valuable sources of information. The Spanish scholar Jiménez de la Espada has collected a number of reports about various provinces from different places and different times which he has titled *Relaciones Geográficas* and which he published in four volumes. Also among the sixteenth century government documents are the *vistitas*, formal questionnaires prepared for some specific purpose by the Spanish authorities, and administered by an official inspector (*visitador*). For the anthropologist and the historian, such documents are gold mines since they were administered in the field with informants who possessed a deep knowledge of traditional Andean life, providing empirical data on specific places at a specific time, among real people, under controlled conditions, much like modern social and economic surveys, yielding data which permit the scholar to make both qualitative and quantitative assessments of daily life in the community surveyed, and describing the nature and function of households and their relationship to their local leader as well as to the Inca state. Those documents have been examined in much the same way as a variety of sources that were analyzed by the Annals School of French historians to provide a more accurate picture of life in medieval and early modern Europe.

Native religious practices are reported in the chronicles and are, along with descriptions of daily life, one of the most informative parts of Poma de Ayala's writing and graphic depictions. In 1579, Father Cristobal de la Molina of Cuzco wrote an account of Inca religion, and in the century after the Conquest missionaries, engaged in a campaign to destroy native religion, give us accounts of traditional Andean beliefs and ceremonies such as Jose Arriaga, the Religiosos Augustinos, and Francisco de Avila who recorded native beliefs in Quechua from the province of Huarochiri.

All of those sources provide data that can be compared cross-culturally with social and cultural institutions from Africa and the Pacific for a better understanding of how such institutions work. And finally the facts given in the historical record can be checked archaeologically as, for example, Craig Morris (1986) has done for the Inca storage and warehousing facilities which are described in the chronicles. With those sources of data anthropologists and historians can develop a clearer picture not only of Inca culture and society, but to a certain extent of the periods in Andean history from which it emerged.

THE INCA PERIOD

Carefully kept genealogies of the lineages of what would become the Inca royal family were transmitted orally and written down by the Spanish. From those records we can trace the Inca rulers back to AD 1200, the first of whom was the Inca Manco Capac. According to those traditions expansion took place from 1438 to 1530 under only three Incas, Pachacuti Inca Yupanqui, Tupac Inca Yupanqui, and Hayna Capac, over just ninety years. The first phase was the conquest of the most closely related polities in Southern Peru and northern Bolivia. The next phase was the conquest and incorporation of Northern Peru, both highlands and coast, as well as the highlands of Ecuador. By the time of the Spanish arrival, virtually all the people that had a state organization or ranked polity, from northern Chile and southwestern Argentina to a small part of what today is Colombia was under Inca rule, thus for the first time politically uniting the entire Central Andean culture area. The Incas divided their empire into four parts with their center in Cuzco, which Garcilaso says was called "the navel of the world," thus the empire's name Tawantinsuyu "made up of four parts:" Collasuyu in the south, Antisuyu in the east, Contisuyu in the west, and Chinchaysuyu in the north.

The question is how could the Incas in that short time have created an empire of such proportions and such a high level of organization that it made the Spanish marvel? Murra says the answer lies in the long tradition of statecraft in the Andes, namely knowledge and practice of how to incorporate and govern different linguistic and ethnic groups and manage the resources derived from different resource zones, knowledge of what he calls the political repertoire of thousands of local lords before AD 1000. The Incas were thus heirs to an experience of statecraft going back centuries if not a millennium (Murra 1986).

Each province conquered by the Inca defended its own territory without help from others therefore the Inca could confront them one at a time without fear of wider alliances. The Inca strategy of conquest was to raise a large force and then confront an ethnic polity, offering its leader a choice of co-operation in an arrangement in which the local *kuraka* would be left in place as a part of the imperial administrative apparatus, or the Inca's superior force would attack. Sometimes the threat of defeat was enough to convince the local ethnic lord to opt for co-operation. Sometimes they resisted and were defeated and forcefully incorporated into the empire. In cases where resistance to the Inca might prove costly or unsuccessful, the Inca would by-pass the area until a large enough territory on its flanks had been secured, and then return with added forces. The more areas conquered, the greater the pool of manpower for future military operations and the greater the resource base for military campaigns. Self interest was one reason why some polities joined the Incas, for they saw the promise of peace under Inca protection in what had for centuries been constant turmoil. Under the *Pax Incaica* violence was managed and redirected to the peripheries of the empire.

The Inca ruled their empire by elaborating ancient traditions. One such elaboration was the tradition of sending colonists from an ethnic nucleus to other resources zones. The Inca elaborated this practice on a pan-Andean scale. Under this system communities loyal to the Inca would be sent to a region which was still restive as a means of pacification. And communities from one resource zone would be resettled in the same zone often quite far away in order to extend the cultivation of maize and coca in the temperate valleys and herding in the high altitude areas. Some valleys, such as Yucay and Cochabamba, were emptied of the native population, which was sent elsewhere, and repopulated with new settlers who worked their own acreage as well as that of the state without obligations to an ethnic lord. Such resettled communities were called *mitmaq*.

Another development was the elaboration of the production, use, and distribution of traditional resources. Maize was of special importance to the Inca state for the maintenance of state officials and for use in the army where it was more highly valued than the ordinary potato. Its keeping quality is good, and thus it could be stored in state warehouses for later distribution. For those reasons and because of its high status (due in part to its relatively restricted growing area), maize was a valuable resource for state activities.

Maize was particularly important in the Inca state religion which included rituals with its propagation and harvest. It was also used in magic rituals. Coca was also important to the Incas who believed it was of divine origin. Under the Inca, coca cultivation was made a state monopoly and *mitmaq* colonists were sent to warm zones for its cultivation. Under Topa Inca, 1471–1493, its use was restricted to the nobility and to those in service to the Inca, such as orators, messengers, and the army where along with maize and *chicha* it constituted part of the rations. It was also important in the Inca ritual process.

The Inca system of governing and managing an empire-wide economy led them to take what was already there and reshape it to their purposes. One important device for those goals was the *mita* or labor tax similar to the medieval European *corvée,* a system which they used to mobilize the labor they needed for both civilian and military purposes. One of the civilian projects for which labor was needed was the construction of infrastructure to bind the vast empire together. Suspension bridges were improved or built and two roads were constructed from the northern to the southern end of the empire, one on the coast and the other in the highlands. In some places the roads were paved with flat stones, and since the Incas had no wheeled vehicles stairs were built on steep grades. Irrigation canals and terraces on the slopes, built with *mita* labor, served to increase arable land.

A system of messengers, recruited by the *mita,* was also initiated with runners called *chaski* carrying oral messages and sometimes objects from one station to another. Rest houses called *tambo* were built along the road a day's journey from one another and storage facilities were constructed in strategic places in the countryside to store the tribute the Inca extracted from the communities. From those facilities the Inca could supply the army on the move and laborers were drafted for state projects. Food from those granaries would also be distributed in time of famine, thus earning for itself in the twentieth century the description of the Inca Empire as a "socialist" state.

All the land within the empire was divided into three parts. One was for the use of the community, another provided resources to support state enterprises, and the third for the state religious organization. At some point the Inca initiated a decimal grid to deal with its relationship with the population as a whole rather than a mosaic of different subordinate groups. In this way they brought a "rational" pattern to the ethnic and kinship based groups of traditional society. The population was divided into groups from ten to one thousand which the state kept track of with officials called *quipucamayo*. This system was ethnically borderless, with some people taken from neighboring ethnic groups when necessary to fill out the grid. It is not clear how extensive this system was, for it did not cover the entire core of the empire in 1532, but it shows a high level of sophistication and rational accounting for state purposes.

The sole point of power was in the person of the Inca. His kinsmen formed the highest tier of imperial society along with those nobles who were designated Inca by privilege. Below them were the *kurakas* which formed a lower nobility, then the rest of the population known as *hatunruna*. There had always been individual retainers to the *kuraka* known as *yanakuna* who had left the *ayllu* and its obligation to serve an ethnic lord. This practice was expanded under the Inca with individuals serving the Inca and the state religious establishment in a number of special occupations such as pottery making, herding, and construction. In some cases this was a hereditary status, the way an occupation is passed down from one generation to another, but in this case accompanied by service to a specific party. Under the Inca a few *yanakuna* enjoyed high status with property of their own and sometimes their own land.

Religion in the Andes was centered on the belief that the objects of the physical world have spiritual essences such as the earth, *pachamama*, maize, *mama sara*, and coca, *cocamama*, as well as the hills, solitary rocks and caves, the *apu*. Manmade objects from piles of stones (*apachita*) to pyramids known as *huacas* were also venerated. The Inca added something new to traditional Andean beliefs with the figure of a Supreme Being which they called Weraqocha. In 1438, Cuzco was attacked by the Chancas. This belief and cult started with one of the leaders who distinguished himself in its defense, Cos Inca Yupanqui, who later became the ruler. After the victory over the Chanca, he conquered the territory around Cuzco thus founding an empire that would rapidly expand. For his achievements he took the name of Pachacuti Inca Yupanqui. (Pachacuti means "cataclysm").

Pachacuti decided that he needed a state religion to provide new and coherent institutional control in order to integrate the diverse peoples of the empire. To provide for this religion he regulated existing cults and established new public ceremonies and sacrifices. What he came up with, however, was not simply a copy of the traditional pattern of Andean belief, but one based on an original and fundamentally different idea. The account of how he devised this religion comes from oral tradition with variations in the chronicles that reported them. According to that tradition Pachacuti was inspired by a vision from which he elaborated the new religion. Its central figure was a Supreme Being, *Teqsi Weraqocha.* This deity had the form of a man, and was considered to be the creator of the world and all that is in it. When his work was done *Weraqocha* departed across the ocean walking across the water. It was later said that the Indians believed at first that the Spanish were *Weraqocha* returning. But that may be a post Conquest legend. The term *Weraqocha* is actually the name of a class of divine figures separate from the Huacas of local religions, with the Supreme Being standing at the top of the hierarchy.

The Sun *Inti* was also a principal Inca deity and the moon, Mama Killa, was his wife. Another major figure in the Inca pantheon was thunder, the god that brought rain. It was said that Pachacuti felt that the sun, a principal deity, could not be a Supreme Being, for the sun could be obscured at any time by any passing cloud. The regular rising and setting of the sun conformed to the laws of nature, and thus the sun could not be the supreme deity, for if it were, it would stand above laws and be able to act of its own volition. John Rowe says of that new vision, and the religious structure built around it, that it "epitomizes the position of the Inca in the Andean cultural tradition, heirs of centuries of earlier efforts which the best minds reinterpreted with creative originality to meet the needs of their successful conquests" (1960: 424).

We can thus regard the Inca Empire and its rapid success in terms of the elaboration of preconditions that had long existed before the lords of Cuzco reached out in their imperial venture to unite a culture area. The limits of the environment were also the limits of the empire, for once Inca expansion involved other social and cultural forms not compatible with the social, economic, and political institutions of the highlands and the coast, the system no longer worked and the empire had attained its natural frontiers. These frontiers in the north were the jungle coast of Ecuador and the Northern highlands where the great mass of the Central Andes breaks up into the separate blocks and are characterized by tropical forest conditions. Julian Steward says that further conquest of the chiefdoms of that region may have been possible, but it was not possible on the eastern slopes where mountain conditions abruptly meet the rain forest of the Amazon Basin. For example, the Incas were never able to conquer the primitive Arawakan and Panoan people of the *montaña* just a few miles east of the Inca capital of Cuzco. The empire reached into southwest

Argentina and as far south in Chile as the Maule River. Conditions on that southern margin were less advantageous for empire, for in that arid region, the driest on earth, water runs off in deep canyons and is thus not available for irrigation as it is on the Peruvian coast. The empire finally ends in the south where the grasslands and forest of Chile begin.

The frontiers of empire in the southeast lay along the margin of the Gran Chaco whose name comes from the Quechua *chaku*, "hunting ground." This is a vast arid semitropical plain which even today is sparsely populated, where the Andean agricultural way of life did not exist and where there were no pre-existent states or chiefdoms to incorporate into a wider political and economic network. In the late fifteenth and the early sixteenth centuries a war-like Tupi-speaking people, known as the Chiriguano, migrated from the rain forest onto the Chaco as far as the flanks of the Andes where they raided the frontiers of the Inca Empire. It is said that as they passed through in the early sixteenth century, they killed and ate some 60,000 Chané who were native to the region. No one knows why they migrated (like some South American aboriginal *Völkerwanderung*), perhaps for new territory, perhaps in search of their ancestors, and maybe even with the prospect of looting the rich Andean culture of which they have heard, but it shows the limits of native Andean civilization on its most highly elaborated and its most closely integrated level.

THE SPANISH CONQUEST

In 1532, Francisco Pizarro with less than two hundred men, some horses, and a few cannons invaded Peru and by 1538 had conquered the entire Inca Empire, the most spectacular military success in world history. William Prescott's narrative (1847) of the Conquest is still worth reading, as is the more recent account by John Hemming. Spanish intrusion could not have come at a more favorable time for the Spanish adventurers and the worst possible time for the Incas, indeed the worst possible time for native Andean civilization. There were four main reasons why. First was smallpox and measles which had been inadvertently introduced to the Caribbean by the Spanish. Those diseases swept in waves of epidemics among the natives who had no immunity, even before the Spanish arrived, demoralizing and debilitating the Andean population. Another effect was the sudden death in 1527 of the emperor Huayna Capac who succumbed to smallpox before having named a successor to the throne, thus setting off a bloody five year war of succession between his sons Huascar and Atahuallpa.

Pizarro and his men arrived just after the final battle of the war in which Atahuallpa had prevailed, only to be captured by Pizarro through deceit in the northern highland town of Cajamarca where he was held for ransom. Gold was sent from all over the empire to reach a mark drawn on the wall of a room by Pizarro. After the ransom was paid the Spanish killed the Inca, setting in motion a revolt of disaffected subject peoples who joined with the Spanish in their march from Cajamarca on Cuzco. In 1534, Lima was established near the sea which provided an important port of entry to the great inland empire and which became its capital. In the beginning the Spanish conquest was fought with an Indian army under Spanish command. Without native manpower the small band of Spaniards could never have accomplished what they did and in such a short time.

Another factor in the Spanish victory was the very different ways they and the Incas waged war. The Inca's motive was territorial gain and their strategy was to amass forces and announce their intention in an effort to intimidate the people they intended to conquer. It appeared to them at first that the Spanish motive was only the acquisition of gold, which had been spectacularly demonstrated by the capture of Atahuallpa and their demands for gold as a ransom. The Spanish, however, hid their broader intentions. European powers would have viewed their pretence with suspicion, but the Indians did not see things the same way, and thus entered into what they thought was an alliance with the Spanish marauders who they believed would advance their own interests.

The conquest period, before final resistance was quelled and Spanish colonial institutions were firmly in place, lasted forty years from 1532 to 1572. During that time the population was reduced by waves of epidemics. Also occasional references to famines indicate the economic disruption caused by the epidemics and the resultant population loss. It has been estimated that the population was reduced by from two thirds to three fourths. In settlements susceptible to Spanish control the *kurakas*, called by the Spanish *caciques*, were put in charge of supervising the cultivation of the land, tax collecting, and enforcing the *mita* thus using the administrative apparatus of the Inca and thereby enlisting the *kurakas* in their own form of colonial government. At that time, a system known as the *encomienda* was introduced which became the most powerful instrument used by the Spanish in the early phase of colonization. Under this system the Crown granted Indian communities to private individuals known as *encomenderos*. The *encomendero*'s duty was to look after his charges and see to their Christianization. His recompense was a portion of their production. In order to prevent the rise of a Spanish petty nobility in the country, grants were made for only a single lifetime, at most three, thus ending with the original *encomendero* or, in some cases, with his grandson. One result was the progressive dissolution of the traditional native communities and a massive swelling of the *yana* class, some of which moved to the cities to serve the Spaniards. In this way an urban and a rural class was created in the native population.

Tribute was collected among the rural population. And alongside communal agriculture and the native economy, there developed a market system associated with European goods. This was also a time of great mobility with landless people on the move and thus the shuffling of once more stable populations. The social spectrum within the native population was narrowing with the hereditary *cacique* class and an expanded *yana* class serving Spanish interests in different ways, and the majority population living in rural communities. The period was also characterized by a civil war between two factions of conquistadores, one led by Pizarro and the other by Diego de Almagro. Pizarro defeated Almagro in 1541. In that same year Pizarro was assassinated by a faction led by Almagro's son who was executed in 1542, but this did not jeopardize the final outcome of the Spanish conquest of Peru.

There was also native resistance in several places. One was known as the *Taki Unquy* (singing sickness) which appeared around 1564. It combined political with religious elements. Its adherents believed that the *huacas* would return and a new order would be instituted, much like the Ghost Dance at the end of the nineteenth century in the United States and the Cargo Cults of New Guinea in the twentieth century; a native response to the shock of massive culture change after contact with the prevailing Western Civilization.

The most notable of those uprisings was the one led by the Inca family and nobility. They and their retinue retreated to Vilcabamba, a remote relatively inaccessible jungle area east of Cuzco, where in 1539 Manco Inca Yupanqui (Manco Capac II) established a small independent state from which he continued resistance against the Spanish conquerors. He was killed in an engagement with the Spanish in 1540 and was succeeded by his son Sayri Tupac who accepted Spanish authority and moved back to Cuzco. Sayri Tupac died there, some believe of poisoning, and in 1571 was succeeded by Tupac Amaru who continued the resistance until he was captured in a final battle of resistance with the Spanish. The Inca was accused of murdering a priest in Vilcabamba and was condemned to death. The Spanish clergy claimed that the viceroy was exceeding his authority and recommended that the Inca be sent to Spain for trial. But Tupac Amaru was executed anyway, on the main plaza of Cuzco to the wails and lamentation of the natives who had come to see the execution of their Inca. King Philip II of Spain also disapproved and eventually recalled the Viceroy Toledo to Spain where he was charged with the illegal execution of the last Inca and also of unlawfully enriching himself. But Tupac Amaru was dead and the final resistance to the new Spanish imperial order was eliminated, although the idea of Inca resistance did not cease until the end of the eighteenth century.

The location of Vilcabama became a mystery, one of those legends of lost cities that have excited the imagination of many in the wilds of South America. Hiram Bingham, when he found the spectacular ruins of Machu

Pichu, thought that he had discovered the site of Vilcabama. The real site, however, is 130 miles east of Cuzco at a place called Espiritu Pampa which was discovered in 1892. Bingham had also surveyed Espiritu Pampa but without recognizing its significance. In the 1960s Gene Savoy and Antonio Santander associated Espiritu Pampa with the legendary Vilcabambas, and subsequent historians and archaeologists have confirmed that it was indeed the place from which the last Incas carried on their last resistance to the conquest of their former empire.

THE COLONIAL PERIOD

After 1572, the conquest was complete, Indian society was relatively uniform and largely divested of its political power, the development of a new colonial order was underway as was the process of Christianization, although the native religion was still intact. This was the time of Francisco Toledo, viceroy from 1568 to 1581 who initiated major administrative changes. Tension was still high from the struggles between Pizarro and Almagro when he took office, and his first years saw the last of the Inca resistance. Toledo toured the region and decided to concentrate on two main points; the state of mining and improving the lot of the native population. He drew up wise codes to govern mining and he took steps to protect the Indians from the *encomenderos*. The *encomiendas* were eventually phased out and were converted to Crown lands which were ruled by royal officials known as *corregidores*. Those officials were policemen, magistrates, and tax collectors with virtual absolute power unchecked by higher authorities. Despite the tyrannical rule of the *corregidores*, the Indian communities were affirmed by the new system.

Another of Toledo's initiatives was the resettlement of the dispersed population, reducing the number of settlements by concentrating them in fewer but larger centers. The most advanced urban planning was employed in that project in which the familiar town square and street grid design was employed in the construction of the new towns and in redesigning the older ones. In the process many smaller, less accessible towns were abandoned.

Another important change was the Spanish adoption of the *mita* as a means principally of recruiting labor for the mining industry which was booming, especially the mining of silver which had been found in abundance. This meant the movement of workers from their homes to a period of labor in the mines. Population movement was also caused by those fleeing the onerous labor draft. Another feature of the period was the increase in the number of priests with the aim of systematic indoctrination and systematically wiping out the native religion. The effect was the kind of syncretism on the folk level that is seen in the contact of old with new religions elsewhere. In the period between 1572 and around 1650 the full work of colonization was in progress.

George Kubler (1946) calls the time between around 1650 and 1750 the Mature Colonial Period since by 1650 the components of the colonial system were in place and mechanisms of colonial rule had been fine-tuned and were running smoothly. This was also a time of relative stability. Economic changes were also taking place. The obligation within each community of meeting subsistence needs and the requirements to pay tribute, to meet the requirements of the *mita*, to pay the priest salaries and church tithes, as well as the modest demand for European commodities, pushed Indian communities toward the development of local products for the market. In this way they became more involved in mercantile activity and the rudiments of a money market. The marketing and transportation of their products became more prominent and old well-traveled trade routes were used and market centers developed.

The period as a whole was characterized by peonage of the majority of the native population, the natives' increased participation in a market economy, and growing burdens. What happened during the late colonial period, from the later half of the eighteenth century, was a gradually worsening situation first appearing in official reports and, in the later part of the century resulting in rebellion. The form those rebellions took, and the nature of its leadership, says Kubler, was likely the result of the union of Enlightenment ideas and nostalgia for Inca times within the affluent India caste. The ideas of Enlightenment philosophers were discussed among Creoles and Europeans in Peru at the time, and a cult of Inca antiquity had long existed within the Indian classes. Ceremonies had always been held in Lima in which members of the Inca dynasty were impersonated in full costume. Also the wearing of Incan clothes was common in Cuzco, as were genealogies and portraits of the dynasty among families claiming Inca descent. Around 1750 this cult assumed greater significance among the *kurakas* (*caciques*) than simple nostalgia and pageantry.

This affluent Indian caste had an independent status in colonial society retaining its hereditary authority and privilege, thereby enabling the Caciques to accumulate wealth and prestige and to acquire servants. They were also loyal to the Crown and to the Church, for its rituals and faith were an important part of their lives. Often the clergy would intercede for them. What they opposed was the *corrigedores* and they were obviously moved by the strengthened cult of the Inca. The rebellions that occurred with frequent intervals throughout Peru in the late colonial period were the usual form of peasant uprisings. They sometimes involved whole provinces but never coalesced into a pan-Andean movement.

The Great Revolt of the 1780s began in the town of T'inta near Cuzco and was led by the *Cacique* Jose Gabriel Condorcanqui who took the name of the last Inca, calling himself Tupac Amaru II. He proclaimed as his goal not independence but rather administrative reforms. He and the force he assembled laid siege to Cuzco in 1781, and at its high point his rebellion controlled the resources of five provinces.

The Tupac Amaru uprising lacked adequate forces and organization and it suffered from poor communication. Nor did it have an appropriate administrative network outside the weak local relationship among local *Caciques*. Eventually Tupac Amaru II was defeated and was taken to the same plaza in Cuzco where the last Inca, Tupac Amaru, had been executed. The second Tupac Amaru was executed there by being drawn and quartered. The horses, however, were not strong enough for the task, and although he died the dismemberment did not occur. A thunderstorm then broke and a legend developed that the slain leader would one day return, a theme seen in other places after the death of a popular figure.

There were two other rebellions in southern Peru at that time. One resulted in the siege of La Paz after the death of Tupac Amaru. None of them was successful, however, partly because they remained separate. If they had united and if they had been successful, says Kubler, they would have established an independent state not one like the neo-Inca state of Vilcabamba, but a colonial style form of government with Inca symbols and nostalgia but with a Spanish philosophy and Spanish forms.

There developed during that period a colonial style of art. Painting was similar to that of other Spanish American countries and Spain during that period focusing on religious themes. Perhaps the best known example is the Cuzco School of religious painting by Indian artists. Kubler describes Andean folk music as the richest and the most original in Native America, a blend of sixteenth-century European polyphony and a native pentatonic scale. Stringed instruments were introduced to supplement the flute, pan pipes, and drums of native music, and the Quechua repertoire of songs includes religious chants, love songs, funeral laments, farewell and pastoral songs, and vocal accompaniments for choreographic and instrumental performances. What is absent are nursery and cradle songs. Today the wayno, with a strong Andean flavor, is popular as well as the European based Peruvian waltz and the marinera. Also in the contemporary musical repertoire are the cumbia and the salsa from the northern part of South America and the Caribbean.

THE ANDEAN LANGUAGES

The linguistic landscape of the Central Andes at the time of Conquest was as diverse as were those of other regions of Native South America. Spanish clergymen reported that there were hundreds of different languages spoken in the Inca Empire, and the *Relaciones Geográficas* with its inventory of provinces shows us the extent of linguistic diversity in the region. This diversity was intensified through the pre-Inca practice by ethnic polities of establishing permanent settlements of kinsmen from a nucleus area to outlying resource zones, where different groups speaking different languages lived in close proximity to one another. As networks of interregional relationships developed so too did channels of communication in which lingua francas emerged to provide a means of communicating across language boundaries. There were three such languages at the time of the Conquest as reported in Spanish documents of the time which the Spanish called *lenguas generals.* They were Puquina, Aymara, and Quechua. In the early sixteenth-century Jesuits were required to learn all three.

Another important language was Mochica (also called Muchic, Yunga, and Chimu) which was the language of the civilizations of the north coast of Peru. Mochica belongs to the Chimuan language family and was first described by linguistic missionaries in 1607. It was widely spoken in the seventeenth and the eighteenth centuries, but lost ground in the nineteenth century and by 1920 it had become extinct and was replaced by Spanish.

When the Incas extended their reach across the width and length of the Central Andes they also extended their own language Quechua (which the Spanish called the *lengua general del inca*), as a lingua franca and as a part of their pan-Andean administrative apparatus. Language spread of this kind is characteristic of imperial and commercial expansion in the ancient Old World, as seen in the case of Aramaic which was spoken from Persia to the Eastern Mediterranean, and with Greek in the Eastern Mediterranean after the conquests of Philip and Alexander the Great, and in the spread of Latin throughout the Roman Empire. We also see it in the overseas colonial empires of Europe with the spread of Spanish, Portuguese, French, and English to the Americas and to Africa.

Quechua was an important unifying element in the Inca administration. The sons of *kurakas* were brought to Cuzco to learn it and to receive an Inca education, and in every province there were people who could speak the language even if for only administrative purposes, and for purposes of wider communication. With the demographic shifts affected by the Inca *mitmaq* policy, an already diverse linguistic landscape became even more diverse, a situation conducive to the spread of a single medium of communication.

The epidemics that reduced the native population in the early years of the Colonial Period, and the movement of people which was characteristic of the time, also contributed to the loss of smaller native languages and the adoption of Quechua. For example, in 1577 Quechua was listed in the *Relaciones Generales* as one of the languages spoken in the province of Chumbivilcas in what is now the Department of Cuzco (which we look at in the next chapter). Aymara was also spoken in *mitmaq* communities in one part of the province, as well as a local language listed as Chumbilcas whose only trace today is its name left in a late sixteenth-century document. Today virtually everyone in Chumbiuilcas speaks Quechua across the range of social classes with Spanish as the primary language of the Mestizo population and Quechua the language of the rural peasant population, many of them, especially women, who know no other language.

The forces that had been set in motion under the Inca in the spread of Quechua were intensified by early colonial conditions and in the late sixteenth and early seventeenth centuries. The Spanish, like their Inca predecessors, employed the *kuraka* (cacique) class as a part of their administrative apparatus, thus helping spread Quechua in commerce and in political affairs. Priests were also involved in spreading Quechua through catechisms and sermons. Philip II approved of those measures, for he wanted the clergy to use the Indian languages in their ministries. He refused to make the learning of Spanish compulsory for the Indian population. In this way royal approval was conferred on what was already underway.

In the sixteenth century a professorship of Quechua was established at San Marcos University in Lima and one in Quito. This policy continued under Philip III, but in 1634 Philip IV ordered the archbishop and the bishops to command priests to use Spanish by "the gentlest means" to spread the use of the language among the Indians. Decrees throughout the rest of the seventeenth and the eighteenth centuries favored Spanish but throughout the period conditions favored the continued spread of Quechua in the highlands (Heath and Lapade 1982, Custred 1992).

One of those conditions was the movement of people escaping the *mita* which led to the break-up of smaller language communities, and the adoption by their speakers of Quechua, a language which enjoyed a high status and which was more useful to a wider range of the population under those conditions. Also the policy of concentration of the native population in *reducciones* often brought together people speaking different languages which also fostered the use of a single, more widely-understood language. As the pool of Quechua speakers grew and as it spread over an ever wider geographic area, Quechua became advantageous for more and more speakers not only of Indians, but also Creoles who lived among the Mestizos.

Quechua's prestige was also enhanced by the clergy who, in the Early Colonial Period, employed it in their efforts to convert the Indians to Christianity, thus further legitimizing its use. As time went on fewer and fewer priests used the other native languages preferring Quechua which by then had acquired a semi-official character. The policies of the *encomenderos* and later the *corregidores* kept the illiterate rural population in isolation and thus without the need to learn Spanish. Thus as a result of colonial conditions and colonial policies we see the extinction of all but a few of the many languages of the Andes and the spread of Quechua, and its exclusive use among the illiterate peasant population in vast parts of the highlands. The situation of the favored cacique class was different for they were a part of the social, political, and commercial system where both languages was useful, and where a special school had been established for them not only to learn literate Spanish and culture, but where they even studied Latin. The caciques were therefore bilingual and literate, and thus involved in both Spanish and traditional Andean culture.

It was also in the later part of the sixteenth and the early seventeenth centuries that we see an example of what Dell Hymes has called anthropological philology, the philology of people who have none of their own. Spanish linguistic missionaries introduced the Latin alphabet, wrote grammars and dictionaries of Quechua, Aymara, and Moche (or Yunka) as well as a sketch of Puquina. They also composed catechisms and manuals for priests with Quechua phrases to help them in their ministry, as well as sermons in Quechua. One priest, Francisco d'Avila, transcribed in Quechua the mythology of one district, Huarochiri, as an aid in countering the native religion, a document which today is a valuable linguistic and ethnographic source of information about the time and the language spoken in that region (Porras Barrenechea 1963: 37–48).

Philological interest in the Andean languages waned during what Kubler calls the mature colonial period; however, for Quechua a new literary interest came into prominence as a part of the Inca revitalization of the eighteenth century. This is seen in a Cuzco school of writers who specialized in popular drama. The most famous work of this period was the play *Ollantay* written by the clergyman Antonio Valdez which, like others of the genre, was based on an ancient legend depicting Indian customs but employing in the Quechua text literary styles of contemporary European origin. Religious literature was also written in Quechua at that time. This literary impulse ended with the quelling of the late eighteenth century rebellions and Quechua, like Aymara and the few remaining native tongues, became a spoken language of low prestige.

Scholarly interest in Quechua among Peruvians picked up in the nineteenth century due its national significance, as well as among foreigners who were motivated by the general scientific inquisitiveness of the time. The American Clement Markham and the Swiss scholar Johann Tschudi directed some of their attention to Quechua. And German physician and long term resident of Peru, Ernst Midendorf, has left us a valuable discussion of Quechua literature as well as descriptions of the Quechua and the Mochica grammars.

Aymara has survived with the same vigor as Quechua and those two speech communities together account for more than half the speakers of native South American languages today. In the late twentieth century Quechua had between seven and eleven million speakers which makes it the most widely spoken of the South American Indian languages. Aymara, spoken in the southern part the Peruvian highlands and the northern part of Bolivia in the Lake Titicaca region, has from a half million to two million speakers which makes it the second most widely spoken South American Indian language.

Quechua is found in different dialects in Peru, Bolivia, and in the Ecuadorian highlands as well as in an enclave in Santiago de Estero in Northwest Argentina. Due to its wider dispersal there are dialect variations. Also because of its geographic dispersion and variability Quechua has not become a factor in ethnic identity as has Aymara with its more compact population and far less internal variation. Four other languages were still spoken in the twentieth century, Kawki and Jaqaru in the Peruvian highlands, which belong to the Jaqí family (of which Aymara is a member), as well as two languages of the Uru-Chipaya family spoken in Bolivia in the Lake Titicaca.

Typologically, Quechua is an agglutinating language with few morphophonemic rules while the Aymara is also an agglutinative language and to some extent polysynthetic. Syntactically they both have verb-final position, and phonologically they both have a series of plain, aspirated, and glottalized stops. Mutual areal influences are clearly seen in both Quechua and Aymara since they share some 25 percent of their vocabulary, but the rest of their lexicons show no similarity whatsoever. This indicates that they belong to two different linguistic families, although some dispute this point.

In 1963, Gary Parker identified two major groups of dialects which he called Quechua A and B. A year later, Alfredo Torero introduced his own classification calling Parker's Quechua A Quechua II, and his Quechua B Quechua I. Torero's classification is now generally accepted.

Quechua I (Parker's Quechua B) is spoken in the highlands of Central Peru in the departments of Ancash, Huánuco, Junín, and Pasco and in a portion of the highlands in the department of Lima as well as in a few small areas of Huancavelica, Ica, and La Libertad. All dialects spoken outside that area are Quechua II which is divided from the Quechua I area by a sharp language boundary. Some believe that the difference is great enough to constitute two different languages within the same family. Torero has labeled the dialects of Quechua II as Quechua IIA, IIB, and IIC. Dialects of Quechua IIA are spoken in Cajamarca and Lambayeque in Northern Peru and in a few mountain villages of the department of Lima. Quechua IIB includes dialects of Ecuador and Colombia and those spoken near Chachapoyas in Northeastern Peru as well as the now extinct coastal variety that was described by Santo Tomas in one of the earliest grammars of the language. Varieties of IIC are spoken in Ayacucho, Huancavelica, also in Cuzco and Puno and parts of Apurimac, Arequipa as well as in Bolivia and parts of Northwest Argentina and Chile. The Quechua-speaking enclave of Santiago del Estero in Northwest Argentina is also a Quechua IIC dialect, brought by Quechua speakers to the region in colonial times.

The greatest area of dialect diversity is in the Central Highlands in the Department of Lima which leads Torero to the conclusion that that is the original homeland of Quechua from whence it spread northward and southward. It became widespread in Inca times in Bolivia and in the last phase of the Inca Empire and during the Colonial Period (as described above) where it replaced other native languages. In its turn, Quechua has been replaced by Spanish in the areas dominated by the Spanish-speaking urban centers of Lima and Arequipa and in most of the Northern Peruvian highlands.

Interest in the native languages emerged again in the twentieth century as they and their histories were increasing studied by scholars. Steps were also taken to develop a norm for use in grammars, dictionaries, and educational material. Both Quechua and Aymara have been designated "national languages," a gesture which officially recognizes their existence and their continued use. And in Peru, Quechua is gaining ideological acceptance as a part of the national heritage.

THE POPULATION TODAY

In the 1820s, independence was imposed in the Central Andes from the outside by Jose de San Martín from Argentina and Simon Bolívar from what would become the republic of Venezuela. Eventually three separate republics emerged in the Central Andes: Peru, Ecuador, and Bolivia. During the nineteenth century the Creoles dominated the political, social, and economic life of those countries. The cacique class was gone and the Indians of the highlands were bound in peonage on estates, many of them quite large and isolated, thus maintaining, many Andean cultural features that go back to before Inca times.

Today the Central Andean countries differ from others in Hispanic South America by the size of the racially unmixed Indian and the Mestizo populations with only a small percentage of the population of European descent, both Creole and the descendants of European immigrants. This stands in contrast to Uruguay, Argentina, and much of Chile, from which the descendents of Europeans dominate. A saying in Argentina goes, "The Mexicans are descended from the Aztecs, the Peruvians from the Incas and the Argentines descended from ships."

There is also a black population of the Central Andean countries of African descent, either pure blooded or mixed with Caucasians (*Mulattoes*) or with Indians (*Zambos*). The Africans were brought to the Andes as slaves by the Spanish in the sixteenth and seventeenth centuries. Their number, however, is very small unlike the high concentration of blacks in northern Brazil and the Caribbean, where Africans were imported to work on the lucrative plantations of the region, while in the Andes there was a large native population to work the land and the mines which formed the basis of their economy. The population of African descent in the Central Andean countries today is concentrated on the south coast of Peru and in the Lima-Callao region as well as on the Ecuadorian coast. In Bolivia they live on the eastern side of the Andes in land once inhabited by jungle tribes, relocated there in the nineteenth century after emancipation. Some also live in the city of Santa Cruz de la Sierra where there are also blacks of Brazilian origin.

There is also a small Asian population of Chinese and Japanese descent, most of whose ancestors came in the same wave of Asian labor migration that came to the United States in the late nineteenth and the twentieth centuries. Peru has the largest number of such people than any other Latin American country consisting of from 3.5 to 5 percent of the population. The other two Andean countries have fewer.

Race, however, is only one factor in the demographic classification of the region. Another is culture, and to a high degree, language. The term "Indian" in this classification is applied to the rural population of indigenous blood with a peasant culture speaking a native language and identifiable by their distinctive dress. Mestizo applies to non-peasants of Indian blood or mixed blood in all social classes who are a part of the national culture and whose language is Spanish or, when bilingual in an Indian language, who identify Spanish as their main language. Another category in the Peruvian highlands is Cholo, people who are bilingual in Spanish and Quechua, who wear a distinctive style of clothing and who are of Indian blood and engaged in commerce.

THE GREAT INTERNAL MIGRATION

Since the Second World War, new highways have been built in the Andean countries, railroads have been improved, air traffic increased and expanded, and port facilities upgraded. Also the radio has improved communication, especially the transistor radio reducing isolation of rural communities and providing a channel for disseminating national culture and thereby fostering an increased sense of national identity. A major part of those changes was a large internal migration which in Peru brought, between 1940 and 2010, some eight million migrants from the highlands to Lima and other cities. There was not enough housing for them in Lima, so the migrants occupied vacant land and built their own dwellings. At first the authorities resisted, but eventually the squatter settlements became permanent residential areas known as *barriadas*. Today the metropolitan area of Lima extends some 125 miles along the narrow desert coast.

A new culture began to take shape among the residents of that new area and others like it, as Mestizo and Indian migrants, many of them illiterate and speaking primarily Indian languages, and their children developed their own forms of co-operation and organization and, as anthropologist Jose Matos Mar argued in 1984, created their "own sense of law and morality." They also created jobs for themselves through an informal economy that bypassed the heavy regulations of the state. They also broke down, or by-passed, political and social structures which had been dominated until then by a small oligarchy.

Matos Mar observes that the Incas of pre-conquest times, the colonial viceroys, and the Creole elites of the independent republic had never created an authentic nation. Instead they had presided over variations of a stratified traditional agrarian society with a small elite at the top and a large isolated rural population at its base. This changed as the society and the economy modernized and with it that mass migration from the isolated rural highlands to the centers of social, economic, and political power. In his book *Another Peru* published in 2009, Matos Mar assessed the situation twenty-five years after his first observations, saying that democracy and the rise of a new social class, which took shape as peasants became citizens, was a process of irreversible transformation. One is reminded by Matos Mar's work of Eugene Weber's account of a similar process of modernization and national integration in his *Peasants into Frenchmen: The Modernization of Rural France 1870–1914*.

Hernando de Soto is a Peruvian economist who has studied in Switzerland and has worked as an economist for the General Agreement on Tariffs and Trade (GATT). He has also served as a managing director of Peru's Central Reserve Bank, director of several Peruvian companies, a member of the United Nations Committee for Development and Planning, and president of the Instituto de Libertad y Democracia in Lima. Like Matos Mar, he examined those transformative processes arguing that Latin America in general is nearing the end of a historical stage similar to the one in Europe when the Mercantile system (which had dominated the European countries from the fifteenth to the nineteenth centuries) gave way to modern social and economic ways of doing things, producing an era of unequaled prosperity and political change. In 1989 he wrote a book titled *The Other Path,* a name taken in opposition to the violent utopian Maoist terrorist organization in Peru which called itself *Sendero Luminoso,* "the shining path." De Soto saw the informal economy which he had observed, and which Matos Mar had described, as a path of internal development that other developing countries could follow.

He regarded this informal economy, working outside the system of complicated state regulation, as a phenomenal and a successful movement. He points to the fact that 90 percent of the public transportation in sprawling Lima at the time was in private hands, some of it organized from the ground up by people who saw the need for more efficient transportation and a way to meet it. The reason for the underground economy he said was that the state, passing new regulations every week, was making the bureaucratic system so complicated that only the largest firms comply, thus forcing the natural entrepreneur within the newly rising class to innovate on its own in order to circumvent the strangling entanglements of the bureaucracy.

Such transformations, however, never come as easily as the history of Europe shows. In Peru the mass migration and the informal and unplanned and unprofessional rise of the *barriadas* has turned the metropolitan areas into ugly, chaotic sprawls, and in Lima, which lies on the desert coast, the availability of water is a growing problem. Moreover the creativity and the dynamism which we saw in the informal economy has reached its limits, with many people now dependent on small subsistence businesses, who are no longer working as active entrepreneurs and aspiring capitalists. Instead they long for more stable employment with regular hours, steady pay, and benefits.

The old political order was indeed changed by the Great Migration and its consequences, yet the core political system is still intact, for no new parties or political institutions have emerged and the restrictive system of laws and regulations that De Soto describes still bears little resemblance to realities on the ground. Also politicians are not pushing reforms which the country needs in the labor market or in politics. The new nation that Matos Mar describes owes nothing to the state, but rather everything to its own efforts, and there is an inherent distrust of politicians. There thus exists today a new division in the long history of divided societies which from the earliest times have characterized the region. Also, in a globalized world, change is a constant. When the commodity boom ended, Peru's economy slowed and with it some of the force that was driving it forward. The country is better off today than in the 1980s but still a long way from a developed country. The processes involved in those transformations and their current status provide a continuing field of investigation for anthropologists.

THE AMAZONIAN FRONTIER

Another significant phenomenon in human history is the occupation and settlement of new land, a process now underway in the Amazon Basin which lies along the eastern flanks of the Andean Highlands. That vast tropical wilderness was once the natural boundaries of Inca expansion and a frontier of Spain's colonial empire and of its successor republics. Conditions there are harsh where mean temperatures are consistently high with diurnal variation between 80 and 90 degrees fahrenheit. Humidity rarely falls below 80 percent made worse by frequent downpours with pronounced seasonal decrease depending on location. Insects of various kinds can be annoying and sometimes fatal. Disease such as malaria, dengue fever, beriberi, yellow fever, and dysentery discouraged large scale settlement, and hostile tribes resisted the penetration of people from the outside.

Transportation was by river, and until relatively recently there were only outposts located along some rivers along with a few farmsteads. This pattern of thin settlement and little influence from core areas represents what has been called "detained societies," regions that Brazilian anthropologist Emilio Willems calls anonymous or unsuccessful frontiers; unsuccessful because they were never fully integrated into their core area, places where the population has adapted to local conditions, often adopting native ways, and in some cases creating hybrid cultures (Willems 1994). This is what happened in Amazonia after the rubber boom of the late nineteenth and the early twentieth centuries leaving some of the workers stranded. In Peru such scattered people are called *ribereños* "river dwellers" and in Brazil *coboclos*. They live in isolated huts and small villages in remote places subsisting on fishing, hunting, small scale agriculture, and trade.

That all began to change in the 1950s, when Brazil, which controls the largest share of Amazonia, began a campaign to settle and develop their portion of that vast still undeveloped territory. In the 1970s, they began an extensive road building program, and in the 1980s the Southern Peripheral Highway was built to the Peruvian border. Dams and hydroelectric power plants were also built in the Brazilian Amazon and Brazilian firms are helping other Amazonian countries in such projects. Brazilians have also financed roads into the jungle in the Central Andean countries from which all countries involved will benefit.

In the 1970s and 1980s the Northern Andean countries of Venezuela and Colombia began to push into their segments of Amazonia. Ecuador has declared itself an Amazonian country with its own colonization projects and the Bolivian government has successfully promoted colonization and development, opening up a fast growing area of agriculture in their part of the jungle. And in the 1990s there were five live frontiers in motion down the Andean slopes in Northern, Central, and Southern Peru.

Recent colonization in Amazonia has been facilitated by diminishing the health hazards of living in the tropical rain forest, through preventative measures and new medicines. Deforestation, however, along with mineral extraction and monoculture in many places (the planting of single crops), have created environmental, economic, and social problems, among them are the effects of development on the indigenous population. All these unfolding events provide a field of study for anthropologists in conjunction with investigators from other disciplines.

There are two cities in the Amazon region on its periphery in the Central Andean countries, Iquitos in Peru and Santa Cruz de la Sierra in Bolivia, in which typical as well as special urban problems present topics for studies in urban anthropology. Iquitos was founded by Jesuit missionaries in 1775. It became a wealthy city during the early twentieth-century rubber boom. Today it is a vibrant modern city of 422, 055 inhabitants with a metropolitan population of 462,783 which makes it Peru's sixth largest metropolitan area. More remarkable is the fact that Iquitos is the only city of its size in the world that cannot be reached by highway, only boat or from the air. The Bolivian city of Santa Cruz de la Sierra is located not in the tropical rain forest but rather in the tropical savanna somewhat to the south in view of the Andes which rise to the east. The city, founded in colonial times, was once a distant neglected outpost as were settlements in the jungle until the middle of the twentieth century. Today it is a growing center of commerce and Bolivia's largest city with a population of 1,364,387.

An interesting area of research for anthropology is how pre-Hispanic peoples dealt with the tropical rain forest environment. That region is known is a culture area characterized by a mobile indigenous populations living along rivers from hunting and gathering and swidden (slash and burn) agriculture. In 1541 and 1542, Conquistador Francisco Orellana and his expedition were the first to travel from the highlands (in Ecuador) to the Atlantic via the rivers of the Amazon Basin. His chaplain, Gaspar Carvajal, kept a journal which was later published. In its pages Carvajal describes densely settled populations along the rivers over extensive cultivated areas, living in fortified towns with well developed roads and monumental constructions. His description was for a long time dismissed as propaganda or fantasy. With increased archaeological research in Amazonia, a picture is emerging for which Carvajal's documentation has become a part of the evidence.

It seems that from AD 800 to 1600, well-planned, large communities in relative proximity to one another did indeed exist. And in the Bolivia jungle earth works were found which made agriculture possible during floods and which were designed to avoid the leaching of nutrients from the soil, a result of the climatic conditions of the area. In post-contact times the population abandoned that way of life and became the semi-nomadic hunters, gatherers, fishermen, and horticulturalists which define the Amazon as a culture area, a case of cultural devolution rather than evolution. This was perhaps the result of the epidemics brought from Europe in the sixteenth century which caused a demographic collapse, leading the smaller groups that survived to the way of life they have had since then, another topic for anthropological investigation.

Another aspect of pre-contact Amazonian culture is what has been called "Amazon dark earth" which is different from the nutrient poor yellow earth which is characteristic of the tropical rain forest. In Brazil it is known as *terra pretado Indio* with indications that it is anthropogenic, for it is found in areas of prehistoric settlement in association with abundant pottery. *Coboclos* today seek it out for their small scale cultivation.

Research is now underway to find an economically viable process to produce these effects for modern agriculture.

In this chapter, we have seen how all four sub-disciplines of anthropology in an integrated fashion, and overlapping with neighboring disciplines, give us a still-developing understanding of one distinctive culture area in one specific region of the world, from the beginning of human habitation to the present.

SIXTEEN

An Andean Ethnographic Experience

As discussed in chapter 10, a set of established practices are followed in ethnographic field work in order to assure the best results. Those features include taking a census of the population, drawing a map of the territory, studying its social organization, its economic activities, its belief system, and examining its political, economic, and social relationship with other segments of the wider society. Requirements for such a task are residence in the community and establishing rapport and effectively interacting with the people. It also means learning and using the native language, or one that is widely used by the people under study.

Despite these standard features, every ethnographic undertaking differs for any number of reasons, such as the type of community studied, the degree of trust on the part of its members of someone from the outside asking unaccustomed questions, and the personality of the ethnographer (his or her "people skills"), factors that are irrelevant in the laboratory and in the field science such as geology, but important in studying communities from the inside. We have briefly looked at examples of ethnographic field work in the American Southwest and on Pacific islands. I now describe my own ethnographic study in a peasant society in the Andean highlands in the high altitude zones of Southern Peru, partly to expand on what was said about the nature of ethnography, and partly to take a closer look at the nature of Andean peasant society in the 1960s, when the field work was done, within the framework of the theoretical approaches to culture and society outlined above.

My first trip to Peru was as a Fulbright student in 1965–1966. I started in Lima where I met Peruvian anthropologists and where I had the opportunity to work for a short time with John Murra who was preparing a mid-sixteenth century *visita* for publication, a brief but valuable apprenticeship which Murra called donkey work but which I saw as an opportunity. My self-appointed task after that was to observe and describe a Quechua-speaking peasant community in the highlands to see how it, and the households that comprise it, operated. Since others were working in the intermediary zones of the highlands, I chose the highest life zone, the *puna,* at and above four thousand meters above sea level (Custred 1977b).

On my first visit to Peru I chose a village which consisted of two adjoined communities Punacancha and Belempata which lay behind the mountain Huanacauri which rises above Cuzco to the south, a mountain which had had mythical significance for the Inca. A road runs from Cuzco to the town of Paruro to the east, a trip by bus or truck of about two and a half hours. I would ride on a truck to the highest point on the route. There I would get off and walk across the *puna* for a couple of miles to the village. I lodged in a typical adobe house with a thatched roof next to the school. Teachers in highland communities come from the Mestizo class and are graduates of teacher training programs at the university. They are thus representatives of the outside world and at the same time parts of the communities in which they work. Most of them in that part of Peru speak Quechua. My experience has been that they are glad to meet someone from the outside with whom they share so much and who are interested in the people with whom they work.

I noticed that llama caravans of some ten or so animals and two men on foot passed through the village from time to time. Once, a passing caravan spent the night in Punacancha. I asked the travelers what they were transporting and they told me it was salt. They had come from the *puna* to a salt mine above Cuzco where they bought salt which they took through the warmer valleys on their way home where they traded it for the maize they could not produce in their high altitude community. This was one clear indication of the kind of relationship between the ecological zones that figures so prominently in the theory of how traditional Andean civilization developed and how it was organized at the time of the Inca expansion. I decided then that I would take that opportunity to study a *puna* community and its relationship with others along the highland gradation of resource zones.

I went to the nearby salt mine where I was given a tour by the manager. I saw the miners at work in the narrow shafts as well as the shrines they had constructed to ward off the evil forces which make such work dangerous. (June Nash has described similar conditions in the tin mines of highland Bolivia.) The manager was very helpful. He gave me access to records of transactions with data on the amounts of salt sold and the location of each buyer. The amount purchased indicated whether the buyers were involved in commercial operations or were peasants from highland communities engaged in the interzonal barter trade. I photographed some of the records and consulted a map to locate the area where some of the peasant traders lived. On the basis of that research I chose for my second trip to Peru the community of Alccavitoria in the province of Chumbivilca some 120 miles by road south of Cuzco.

My second stay in Peru (1969–1970) was sponsored by the Midwestern Universities Consortium funded by the Ford Foundation. At that time the consortium was associated with the Agrarian University in Lima. With credentials from my home institution I received credentials from the Agrarian University and with those I obtained an *oficio* from the Peruvian authorities, a letter to local officials stating that my research was authorized by the government. At that time there was much interest in economic development and both the Peruvian government and a number of agencies from other governments (the United States, Europe, and Japan) were working together on several development projects. The Peace Corps was also active, and their purpose and presence was well known throughout the country. Entry into the region, and into the community, therefore, was no problem. In fact my ties with formal institutions were an asset both to me and as it turned out to the community.

Alccavitoria is not a clustered village, but consists of households scattered across the territory, as is usual for pastoral communities at that altitude. There was, however, a nucleus of a school, a chapel, and a couple of houses. I lodged in a small house near the school and took my meals with the teacher who was glad to have the company, as indeed I was glad to have his. Chumbivilcas is remote. Routes from there and from Arequipa, Peru's second largest city, merged to form a single gravel road which at times forded streams, and that terminates near the center of the province at the small town of Santo Tomas which serves as the provincial capital. Traffic from both Cuzco and Arequipa entered Santa Tomas only once a week. It consisted of a bus from each city and trucks which carry merchandise and merchants to the weekly Sunday market in Santo Tomas. On their way home, the trucks stopped at a bend in the road called Esquina for a Monday morning market. During my stay in Alccavitoria I made several trips to Cuzco. On Saturdays I would ride my horse to Esquina, a ride of about an hour, where I took the bus to Santo Tomas. On Monday I would then travel to Cuzco. Before returning, I would go to a local radio station and, for a small fee would have a message sent to the teacher on a daily radio program, a service designed for the purpose of sending personal messages within the broadcast area, asking that my horse be sent to Esquina on the following Monday.

Knowing the local language is an essential part of any ethnographic field work. I knew Spanish long before I went to Peru, and so that was no problem. Some people in Alccavitoria also knew Spanish, but the language most spoken, as elsewhere in the rural Southern highlands of Peru, is Quechua. I began my study of Quechua in graduate school in a field methods course offered by the linguistics department and supervised by Professor Wolfgang Wölck. Our informant was a fellow student, Pedro Plaza from Bolivia, who later became a linguist in his home country. My topic in that class was non-terminal derivational suffixes of the Quechua verb, but I also had the opportunity to work on lexical semantics which would be of use in my later field work. This gave me the opportunity both to learn linguistic field methods and to become better acquainted with the language and with the elicitation of linguistic data. I also took a field methods course with visiting professor Martha Hardman, using data (unfortunately without a live informant) from the Aymara language which also was useful in developing my skills in field linguistics. In Lima, on my first trip, I took a short course in Quechua and worked briefly with a native Quechua-speaking informant.

My use of Quechua in the field, however, was limited, for I never attained the same level of fluency in that language that I have in Spanish. Much of what I learned about the life and thoughts of the people was in Spanish from bilingual residents. My major source of information on the kinship system however was from conversations in Quechua with a very willing elderly woman, Luisa, who patiently went through her family relations with me. From those sessions I learned the lexical structure of the domain of kinship, which like so much else in rural native Andean life is a merger in some ways between a purely Andean and the imported Spanish system. From her I also learned how kinship ties were used in labor mobilization (Custred 1977a). She also told me about her relationship with the church which gave me a better idea of the synthesis of Catholicism, which they all profess, and the native folk religion which goes back to pre-Inca times.

There were other occasions when I gathered information about folk beliefs from informal conversations in Quechua. One was when I went to a river crossing with an old man. I inadvertently touched a cactus and he chuckled at my clumsiness as he removed the thorns from my fingers. While he was doing that he told me that a *supay* dwelt in the river crossing, a malignant presence that the Spanish linguistic missionaries translated as "devils." People knew there was a *supay* there, said my informant, for the river current sometimes grew strong and might sweep travelers away as they forded the stream. Another time a woman told me that there was a *supay* in the lake where we were standing, because two children had drowned there. There are also legends of supernatural and bizarre figures on record. The people I spoke with were acquainted with those legends, and it was from those sources, and conversation like the ones described above, that I pieced together a picture of their folk beliefs (Custred 1979b).

One of the basic tasks in ethnographic field work is taking a census. This involves gathering information on the age, names, and the marital status of the population as well as information on household size and composition. Another basic part of research is making a map of the territory and eliciting information on land tenure and land use. Here I was fortunate since the Peruvian Department of Agriculture had prepared a census with questions on all relevant categories, both on the demographic composition of the community and the assets of households. Here is where my connection with formal institutions and the nature of my task coincided with the interests of the community.

A provision of the law gives advantages to properly recognized indigenous peasant communities (*comunidades indígenas*). The teacher asked if I could help in the application for such recognition. We discussed it with village leaders and the project was begun. The application included a population and resource census, a map of the community, and a valid land title. Their land title was a copy (the original was in the archives in Cuzco) of a seventeenth century document prepared by a priest, Father Cabrera, a survey that had been required by the Spanish colonial government as a part of an inventory of the district. The document described the boundaries of the territory at that time which accurately described the community's present boundaries.

Together with the teacher and a party of men from the community we set out on horseback to map the boundaries as described in the document, what surveyors call a route traverse. The points named by Father Cabrera still bear the names they had back then. We went from one to the other where I located each on a map. The map was based on aerial photographs and published in Lima by the military institute of geography. My method of locating those points on the map was by means of triangulation using a compass and an altimeter and prominent mountain peaks as points of reference. At the start of the trip the men from the community burned an offering and poured a libation to the spirits of the hills. They did the same at each spot that was considered a *saywa* or place of ritual significance. I later took the sketch to a draftsman in Cuzco who drew a proper map of the community.

The census was written in Spanish; therefore the teacher and I had to translate the questions into Quechua. In doing so we encountered some problems, for the categories underlying standard census questions did not always correspond with the corresponding folk categories in Quechua. For example, people do not think about their age in years, even though Quechua has a decimal number system which they use in everyday economic activities. Many respondents, therefore, could not answer the question "how old are you?" One young man answered "ninety." The teacher and a couple of others shook their heads and said, "You can't be ninety, Gonzales [the oldest man in the community] is not even ninety." Here ethnosemantic analysis proved useful.

Since age, reckoned in the number of years, is necessary for a census and for demographic analysis, another way had to be found to answer the required questions. This we did by eliciting names for different stages in the life cycle; *wawa* "baby," *herque* "young child," *sipas* and *maqta* "girl" and "boy" from puberty to young adulthood, *warmi* and *qhari*, adult "woman" and "man" respectively, and *machu* and *paya* "elderly man" and "elderly woman." We thus translated questions on age in those terms and asked how many of each category was in a family. Then estimating the age range of each category we had the rough basis for answers to census questions related to age.

Other questions dealt with the kinds of crops they grew and the number and kind of livestock owned by each family. Those questions presented no problems. Questions about land tenure, however, ran into the same difficulty as did questions about age, since the census asked how much land, measured in hectares, each family owned, yet the people there measured land units not in terms of hectares but rather in terms of *masas*, how much land can be plowed in a single day by two men with *chakitaqllas*, "foot plows" and a woman turning over the loosened turf.

Most of Alccavitoria lies above the upper limits of agriculture, therefore herding is their main occupation. There is, however, enough land available for tuber cultivation, the only crop possible at that altitude, and even then there are years when the frost destroys the crop. Arable land is divided into *muyus* or sectors. Only one sector a year is cultivated, leaving the others to lie fallow. Not everyone has arable land, and some families have land only in some sectors. The question "How many hectares of land do you cultivate?" had to be translated into "How many *masas* do you cultivate in each *muyu*?" The next task was to estimate the number of hectares belonging to the family. Hectares have a fixed numerical value, while the *masa* is variable not only because of differences in output of workers, but also because of the nature of the land cultivated. Some owners showed me sample *masas*, some on sloping land, some on flat places, some with rocky soil, and some with better textured soil than others. I measured each sample in square meters and converted them into hectares for a figure representing their total land holding in all sectors (Custred 1980b).

The project meant some trips to Cuzco as well as my time in completing the route traverse and the census. When the teacher explained the project to community leaders, they asked how much I was going to charge them. It is not always appropriate to simply say that you are going to do something for nothing, especially in a society where people are suspicious of apparently altruistic acts. The teacher watched as I considered the matter. I then said that since the university was paying for this project, I didn't think it would be fair to charge them for it as well. The teacher smiled, the others nodded and we went on from there. The census part of the project was a somewhat formal affair. Heads of households lined up and the teacher asked the questions while I wrote down the answers. I could imagine another foreigner, Father Cabrera, asking the same kinds of question and recording similar answers from the ancestors of the residents of the same community three hundred years before.

The next step was a sketch of the boundaries of the community. When we went on the route traverse on horseback, a party of community members accompanied us on one side of the territory. When we arrived at the Velille River, which divides the community's territory in two parts, another party of horsemen from the other side met us in midstream with a Peruvian flag, and to the music of Andean flutes we rode to the other side together. An example of the kinds of tensions that can arise among neighboring peasant communities occurred when we reached the end of the journey. This was a particularly productive strip of arable land, the kind of land where the boundary between two communities was often disputed. One of the community leaders in our company said that if the other side offered us a cola drink or beer out of bottles that we could open ourselves, we should accept them. If *chicha* (a homemade brew) or *trago,* a raw rum made from sugar cane, were offered in open containers, we should reject them for fear of poisoning. This warning was probably hyperbole, but it did indicate that there were sometimes tensions and animosities among neighboring communities, an interesting line of inquiry that I did not have time to follow up on. Of course the land title, based on a seventeenth century document, was not precise enough to clarify the exact boundary line, and since it was of no importance to me or to the Peruvian government, I asked them where they thought the boundary was, then paced off a line that satisfied both parties.

This attention to detail in translating the census and in drawing the map of community boundaries went beyond what was expected by the Ministry of Agriculture, which understood the problems involved in such questionnaires and boundary maps, but the process was useful for ethnographic purposes. The community was also pleased with the results, for it gave more authority to their permanent records. Some participants later showed their appreciation.

Rapport with the people one is studying is important in ethnography. In my case rapport came partly through the authority of the state and the university and the general feeling that such research somehow had its proper place, as well as from the acceptance of my project by the teacher and its incorporation into a community endeavor. That tie was apparently strengthened on the last night of the route traverse when just after sunset, while there was still some light in the sky, we came to a lake that lies well over four thousand meters in elevation. One of the men shot some water fowl. He rode out on his horse in the shallow water to retrieve the game.

We then went to a low circular shaped building made of stone with a thatch roof in a place nearby called Waykurutu. It was cold. They brought me hot tea laced with raw rum. I drank a couple cups as they were offered. The men brought out their flutes and in the flickering candlelight, in their ponchos and distinctive Andean knitted caps, they played traditional Andean music. I got up to join them in a dance. I suddenly found myself on the floor looking at the dry earth. It was the altitude, for we were well over four thousand meters, and that sudden movement proved a bit too much. That's not how they interpreted it. They never really understood why I sometimes stopped for breath when I was trying to keep up with them on a brisk up-hill walk. A few times after that with a twinkle in his eye, someone would say, "Hey, do you remember Waykurutu?" just another guy, doing a job like everybody else, having a little fun after a long day's work, somebody with whom you can comfortably cooperate.

Written material in this predominately monolingual and non-literate community is still important since the community is a part of a broader whole which depends on writing and literacy for many transactions. For example, Luisa, my informant on kinship, spoke only Quechua and could not read. She once asked me to look over her papers. She kept them tied up in a cloth. They were the kind of household documents one might expect, among them receipts, some for masses, some for family members. They provided the opportunity to talk about the peoples' relationship with the church and their belief in saints, along with the native beliefs in the spirits of the earth and the hills.

There was one family in the community that was esteemed outside the community and could be rightly considered Mestizo. Its oldest member told me that his family had once carried on a profitable wool business, transporting locally produced wool by mules to Arequipa to the south. He said they had kept accurate records but since he saw no use for them he threw them away. He was as disappointed as I when I told him that they would have made an excellent source of information on commerce before the construction of roads. There is no doubt that other such documents are found in what one would consider unlikely places which could augment the ethnographic and historical record, a point that Frank Salomon and Mercedes Niño-Murcia in their ethnography of writing in a rural highland community (2011).

What drew me to Alccavitoria was the practice of trading salt for maize and potatoes in the middle elevation zones, a rational strategy in a community where only meat, wool, and some tubers can be produced locally, and where both the market economy and barter functioned side by side. This strategy also represents a continuity of the flow of key resources up and down the varied Andean environment which on a larger more elaborated scale was an important aspect of ancient Andean civilization. I had hoped to accompany a llama caravan on such a journey but was never able to. Instead I interviewed men who told me about trips they had made over the last few years (Custred 1974).

The process begins with sheering sheep and alpacas, then selling the wool on the local market for cash. They then take the money they have earned to the salt mine above Cuzco where they buy the salt and trade it in the warmer valleys for the resources they need, but which they cannot not produce themselves. From those interviews emerged a clear pattern of a significant economic activity, with assessment of costs and benefits and rational decision-making, using both the market and money and barter to maximize the use of scarce resources in order to procure a necessary commodity, a strategy which reveals one way that peasant communities adapt to the resource limitations of the highest life zone in the highlands.

Since then others, working in the high altitude zones of Bolivia, have described such patterns of exchange based on the salt trade (Lecoq 1987 and 1988; Molina Rivera 1987; West 1981). Ethnographer Axel Nielsen made such a trip with caravan from which he produced a detailed description of routes, provisions, personnel, equipment, camps, trading strategies, and rituals performed while underway (Nielsen 2001); all information useful for a wider ethnographic understanding of native Andean economic strategies, and useful in interpreting the historical and the archaeological record of the indigenous South Central Andean highlands.

Continuities from ancient times, of which the caravan trades described above is one example, are readily observable and sometimes striking in the everyday life and activity of the people; from weaving cloth, to the way people carry goods on their backs, plowing with the traditional foot plow, and so forth. In the seventeenth century Guaman Poma de Ayala (previous chapter) wrote a valuable document in which he drew pictures of everyday life in his time which match photographs of modern peasant life. Continuity is also seen in the beliefs and practices of their folk religion which merges with orthodox beliefs of the church. This is also seen in the language and in the belief system on the folk level.

Tosi has identified thirty-seven different vertical life zones in the central Andean highlands (previous chapter), yet in Quechua this is reduced to only three, *puna*, *qheshwa*, and *yunka*, each defined by what they produce of highest value for the population. When asked what each word means, the answer was that the *yunka* is the zone where coca grows, *qheshwa* where maize grows, and *puna* where no crops are possible, but where sheep, alpacas, and llamas are herded. When asked if there are more terms in that domain, the answer is "no." Alccavitoria does have some land where tubers are grown which otherwise resembles the *puna*. When asked if that too were *puna* the answer is, "No, it can't be *puna* because potatoes grow there." When I asked one informant what it's called he said *pampa*, or "flat area." When asked if there were pampas in the *puna*, he said "yes," and in the *qheshwa* and *yunka* as well? Again the answer was "yes." The next day he came by and said, "You know, we don't have a word for it."

Continuities associated with the vertically layered mountain environment are seen not only in the vocabulary, but also in the symbolic system of the rural people of the highlands where such vertical layering is still important. This reflects today the ancient emphasis on the basic native Andean resources of maize, coca, meat, and wool. When we went on our route traverse community members in the party burned offerings at those places designated in the land title as ritually significant, and before traders go on trading trips with their llamas and when they return home, the same ritual is performed along with the pouring of libations. Those offerings are also made when building houses, and on special days in the year dedicated to the animals they herd; Carnival time in February for the llama and the alpaca, Saint John's day in June for the sheep, and in August "when the earth comes alive" for the cow, a ceremony which they call *t'inkana* which means to give something that acts as a boost or a stimulus to some action (Custred 1980a, 1979a).

The source of each of the major traditional resources, wool and meat from llamas and alpacas, maize, and coca, has what might be best described a "species soul." The species soul of llamas and alpacas is *paqomama,* of maize *saramama* (*sara* means maize), and of coca *cocamama*. Besides those traditional resources, sheep and cattle, both imported from Spain and of high value, also have species souls, *ch'apimama* for sheep and *t'anumama* for cattle. *T'anu* means "soft and silky." When asked why the name *t'anumama,* one informant said, "Haven't you ever felt a cow's nose?"

There is no species soul associated with the potato, just as there is no word for the zones where tubers grows. In Alccavitoria there are also no special rituals associated with tuber and cereal cultivation (quinoa and barley) although rituals associated with tuber cultivation in pre-conquest times were practiced in high altitude zones, but apparently without a belief in a "tuber mother." One reason why tubers are left out of that nomenclature and ritual cycle may be that tubers are so plentiful and found in such variety and in the highland area, forming a basic staple in the Andean diet from very early times. Coca, maize, and animal products were more restricted in their distribution and thus the center of traditional exchange with added social and ritual value.

Only families with those animals perform the rites by pouring libations to the earth then burn an offering (also called *alcanzo*). The offering consists of three pieces of dried animal fat, three kernels each of yellow, white, and purple maize, and three bunches of coca leaf with three leaves in each bunch together with church incense laid out on sea shells. When burned, the offering gives off thick smoke and a sweet aroma. In this form the offering is received by the *aukis,* the spiritual essences of the hills. One informant said that the *aukis,* like us, have to eat and if the offering is not made the *aukis* might take sustenance from the *sonqo,* "heart, life force" of the animals, thus sickening them and diminishing their usefulness. This practice parallels the traditional payment of tribute to higher authorities in the social sphere. The ritual also goes back before the time of the Inca, for it was well established in the early colonial period as described in a manual for priests written in 1631 by Father Pérez Bocanegra, as a part of his instructions to parish priests on how to question Indians on their native practices for confession. In this case the question was

> When offering potatoes, picking and husking corn, when harvesting and threshing quinoa, when dividing llamas and shelling corn, and when building, have you worshipped and made offerings with sheep, llamas, peppers, fat, coca, chicha, red shells and then white shells, gold, cloth, feathers, or any other thing?

The image of mother in the major resources parallels the term *pachamama* the spiritual essence of the earth. Since tubers are cultivated on part of the community territory there is ritual recognition *pachamama*. Each time someone drinks trago or chicha, a small libation is routinely poured to her even on ordinary occasions. Once I asked what *pachamama* was like. The answer was, "she is our mother. She gives us to eat. We lie down upon her breast to sleep." This affective element is also seen in the songs which are sung in rituals dedicated to each of the spiritual essences of the major resources. Since no maize and coca are grown in Alccavitoria there is no ritual for *saramama* and *cocamania*, but there is for *paqomama* for the spiritual personification of llamas and alpacas, and for sheep *Ch'apimama* and for the cow *t'anumama*.

Only women sing such songs and only on the designated day. I asked if someone would sing them for me, but nobody wanted to out of season. Then, while standing around one afternoon talking about it, a woman who had had a few glasses of *aguardiente* suddenly burst out in song. I quickly turned on the tape recorder, and to the sound of trout frying in a pan nearby, she sang for me the song for *t'anumama*.

> Pataykimanta qhawarqamuyki
> Sapachallaykin waqasharkanki
> PobrePeruana
> Hispañamanta lloqsirmurqani
> Pobre peruanata uywasaqnispa
>
> Ay, Ay, señora
> I look at you from your place above
> You were all by yourself alone and crying
> Poor Peruvian woman
> I have come from Spain saying
> I will nourish the poor Peruvian woman
> Ay, Ay, Señora

The song was sung to a tune typical of native Andean music, the text was foregrounded by devices on the prosodic and on the lexical levels. On the prosodic level the first, second, fourth, and fifth lines consist of two words each, each word with five syllables with the accent falling on the penultimate syllable. The fifth line contains four words but has the same metric pattern as the others since the accent of the line is on the fourth and ninth syllable. The third and the sixth lines both have five syllables with equal stress on each, emphasized by the beat of the melody. The poetic devices that foreground the text are parallelism and contrast, with the repetition of the length and metric pattern of the first and second lines, which contrast with the length and metric pattern of the last line of the stanza, repeated in the other stanzas of the song.

x x x /x x x x /x

x x x /x x x x /x

/ / / / /

x x x /x x x x /x

x x x /x x x x /x

/ / / / /

Figure 16.1. Metric Pattern of an Andean Ritual Song

The diction of the song emphasizes emotion; "you were all by yourself along and crying," "poor Peruvian woman," "nourish the poor Peruvian woman," "ay, ay señora." The word *sapachallaykin'*, which I have translated as "all by yourself alone" is literally "your all aloneness" emphasized with the suffix *–cha*. "Crying" is also used in everyday language in a metaphoric sense to denote sorry, as in "Your mother misses you. She is crying." Here we see how the grammar and the expressions of ordinary language have been marshaled for poetic effect in order to enhance a particular event. It is clear from the song that the roles of the woman and the cow are reversed as seen in the phrases "I look at you from your place above" and "I have come from Spain, saying I will nourish the poor Peruvian woman."

Continuity of belief is seen in syncretism with the beliefs and rituals of colonial Spain. Throughout Latin America, communities and occupations have their own patron saints, and in rural communities the saint's day is the occasion of a public festival with a Mass said by the priest, music (in the case of the Indian communities by a mestizo band), and a procession. In Alccavitoria the saint is Our Lady of the Rosary. Such festivals serve on the social level to emphasize community solidarity and its tie with the broader society through church participation and the participation of other groups within a ritual context. There is also continuity here from ancient times, for each community in pre-contact times had its *huaca* and accompanying festivals, a place now filled by the community's patron saint.

The most striking example of this syncretism and wider ritual networks is the Corpus Cristi celebration held in Cuzco in June sixty days after Easter Sunday in which all strata of society, including nearby peasant communities, participate. The night before the procession special foods are prepared. Among the traditional drinks consumed is *chicha* which also played a part in ancient rituals. The next day bells are rung. Among them is the oldest bell in the city, called María Angora, which was cast in the sixteenth century. The central point of the ceremony is the Cathedral on the Plaza de Armas, the main square in Cuzco, which was built on the site of the Inca Supreme Being Teqsa Wiracocha. The procession begins at 11:00 a.m. The statues of fifteen saints and virgins are brought to the Plaza de Armas by representatives of different associations called *cofradías* that carry the statues from different directions to the cathedral (I saw some men from Punacancha there bearing a statute of their patron saint), as had been the mummies of noblemen, priests, and Inca governors. The statues are brought there to meet the body of Christ which is represented by the image known as Nuestro Señor de Temblores, or our Lord of the Earthquakes. The image wears a crown made of flowers called *ñucchi* which were used as offerings to Weraqocha. When Nuestro Señor de Temblores is brought before the cathedral and is rocked back and forth in front of the crowd on the plaza, some people might fall into a fit of religious ecstasy. The statues are then placed in the cathedral and community leaders meet to discuss local affairs.

In this chapter we have described one ethnographic experience to illustrate how this kind of work is done. It has also given us the opportunity to see how the contemporary life of a traditional Andean community reflects long term continuities from the past. All communities change, and as bridges are built between the rural highlands and the growing urban centers of the country, the traditional ways of life described by the ethnographer will become part of a historical record in a continuum from the past into the future in order to allow us a fuller understanding and appreciation not only of a distinctive culture area, but also of the human experience in its wide range of variation.

Epilogue

A Holistic Science

In the preceding chapters, we have seen how anthropology constitutes a branch of scientific inquiry, and that the holistic approach is not only feasible, but is also consistent with the natural process of scientific research. We have also noted that the scientific holistic approach has been challenged by increasing specialization within the four subdisciplines as well as by demands that anthropology abandon its mission as a scientific enterprise and instead become an instrument of social action.

This, however, is a false dichotomy, for both scientific anthropology and activism can be, and have historically been, carried on without the one supplanting the other. In fact the two have often been mutually beneficial, as when social problems can stimulate new thinking about biological, social, and cultural reality, leading to new lines of research and greater understanding, and when scientifically derived and confirmed data can support sound public policy as well as social reform.

The fact is, political opinions and activism have never been far from anthropology. In this regard Thomas Eriksen and Finn Nielsen in their history of the dicipline have observed that "anthropologists have always been a radical bunch" (2000: 111). We can see this as far back as James Adair and John Heckewelder before modern anthropology even took shape. Both were intent on correcting the image that Americans and Europeans had of the American Indian. John Wesley Powell has been described as a fearless defender of Indian rights and an advocate for reservations. And pioneer ethnographer Frank Cushing was forced to leave Zuni for opposing attempts to take it over by United States Army officers. James Mooney was born in Indiana of Irish ancestry but was passionately devoted to Irish causes, promoting Irish interests in the United States and in Ireland. In 1900 at a memorial service for Daniel Garrison Brinton, Albert H. Smyth said in a keynote address at the American Philosophical Society that "In Europe and America, he [Brinton] sought the society of anarchists and mingled sometimes with the malcontents of the world that he might appreciate their grievances, and weigh their propositions for reform and change."

In Britain, writes Robert Lowie, Edward Tylor "Like other evolutionists . . . sometimes felt he had a key to the law of progress" for he "confidentially spoke of anthropology as 'essentially a reformer's science,' as 'active at once in aiding progress and removing hindrance.'" But, says Robert Lowie, "such subjectivism is not at all obtrusive in his writing" (Lowie 1937: 85). Max Gluckman was left-leaning, and the Manchester School which he founded was known for its political radicalism. Peter Worsely, one of its members and author of *The Trumpet Shell* and studies of cargo cults in Melanesia, was a card carrying Communist, a fact which caused him some difficulty in finding employment. And both Durkheim and Mauss were socialistically inclined.

Franz Boas was a strong lifelong proponent of science. He was named a member of the National Academy of Science in 1900, the New York Academy of Science in 1910, and in 1931 the American Association for the Advancement of Science. He was also awarded honorary membership in scientific societies of several countries during his lifetime. At the same time he was also deeply committed to the principles of the 1848 Revolution in Europe, known in the German lands as the *Völkerfrühling*—"the People's Spring." That revolutionary movement was waged on behalf of populist claims to political power against the authority of the traditional dynastic governments of the time, with demands for liberal-constitutional governments and freedom of the press and an end of censorship.

Boas's family had participated in the 1848 Revolution, and some of the participants of that conflict came to the United States. It was in such liberal circles that Boas grew up in Germany and with which he associated when he came to the United States. During his professional career Boas opposed the concept of racial superiority in both the North and the South and was an opponent of eugenics. He is credited as the first scientist to publish on the topic of racial equality. He also told scientists that it was their responsibility to use science to argue against myths of inherent racial superiority. In 1906, he was invited by the premier black intellectual of the time, W. E. B. Du Bois to give the commencement address at Atlanta University where he told black graduating students not to look to whites for approval, for it takes a long time for people in power to sympathize with those who have no power. He pointed to the innovations and the riches of traditional African culture to show that the inferior status blacks had in America could not be explained by Africa.

Boas was also a participant in German-American life. He served as secretary of the German Society of America and after the First World War he was active in the Emergency Society for German and American Science, serving as its president from 1920 to 1928. Although an American citizen for twenty-eight years, he was pro-German against the French-British alliance in World War I, a conflict of national rather than of ideology interests. And like many other Americans he was opposed to American participation in that European conflict. In that matter, as in others, Boas "spoke his mind according to his convictions" but "always within the limits of correctness, spoke it publically and repeatedly, and after 1917 as well as before, unmindful of what the public storm he provoked and the troubles he invited" (Kroeber 1943: 19).

One venue for airing his views was the *New York Times*. In a letter he wrote to the *Times* in 1916, before the United States entered the First World War, Boas protested the bitter attacks hurled against the then large and visible German American community. He was not motivated by ethnic sentiments but by his opposition to ethnic animosity on the part of the majority against a minority. Boas further "spoke his mind according to his convictions" after the war despite the public storm that he invited with his frankness. This time criticism came from within scientific circles. In 1919, he wrote a letter to *The Nation* (a successor to abolitionist William Lloyd Garret's *The Liberator* and the self-described "flagship of the left") protesting in the name of science the actions of four anthropologists, unnamed in the letter, who had combined intelligence work for the government, with research in Mexico during the war. This led one of his rivals, H. Holmes director of the Field Museum, to write one of his friends with ethnic invective, complaining about "the Prussian control" and the need to end Boas's "Hun regime."

The Anthropology Society of Washington passed a resolution condemning Boas's letter and its criticism of President Woodrow Wilson. They also accused him of endangering anthropologists abroad. The resolution was passed on to the AAA and the National Research Council, and at a meeting of members of the AAA at the Peabody Museum of Archeology and Ethnology at Harvard, the committee voted to censure him. As a result Boas resigned as the AAA's representative to the National Research Council, but retained his membership in the AAA, an organization of which he had been president from 1907 to 1909. The resolution was not rescinded until 2005 (see Darnell et al. 2015).

In all of those cases, and there are many more, anthropologists have acted on their convictions in behalf of a number of causes. Yet, until now, they have kept their activism separate from their scholarly pursuits. Those in the discipline today who would extirpate the scientific and holistic approach in favor of some version of political action are flaunting a tradition in which the individual is free to choose his or her political position, thus imposing a single narrow regime on the discipline as a whole. In this way they are both diminishing anthropology and damaging its credibility, a field of knowledge that is both a legitimate branch of science and of great public interest and public benefit. Knowledge of the sound philosophical roots of American anthropology has never been widespread, and under pressure from activists, awareness of its history as a holistic science, and the appreciation of what it has achieved, has faded even more. It is hoped that this short history of anthropology as a holistic science can continue, or revive and strengthen the knowledge and appreciation of a discipline which still has much more to offer.

Bibliography

Al-Khalili, Jim and Johnjoe McFadden. 2015. *Life on the Edge: The Coming Age of Quantum Biology*. Bantam Press.

Arlotto, Anthony. 1972. *Introduction to Historical Linguistics*. Boston: Houghton Miffin Company.

Bailey, Frederick. 1969. *Stratagems and Spoils: A Social Anthropology of Politics*. New York: Shocken Books, Inc.

Barnhart, Robert K. 1988. *The Barnhart Concise Dictionary of Etymology*. HarperCollins Publishers.

Barth, Fredrik. 1969. *Political Leadership among the Swat Pathans*. London: Athlone Press.

Barth, Fredrik. 2005. "Britain and the Commonwealth." In *One Discipline, Four Ways: British, German, French, and American Anthropology*. The Halle Lectures. Chicago: University of Chicago Press.

Bartusiak, Marcia. 2015. *Black Hole: How an Idea Abandoned by Newtonians, Hated by Einstein and Gambled on by Hawking Became Loved*. New Haven, CT: Yale University Press.

Bates, Daniel. 2005. *Human Adaptation: Ecology, Government and Politics*. New York: Pearson.

Bates, David G. 2005. *Human Adaptive Strategies: Ecology, Culture and Politics*. 3rd edition. Boston: Pearson.

Bauer, Henry. 1992. *Scientific Literacy: the Myth of the Scientific Method*. Urbana, IL: The University of Illinois Press.

Berkhofer, Robert. 1968. "Introduction" to James Adair, *The History of the American Indian Particularly Those Nations Adjoining to the Mississippi, East and West Florida, South and North Carolina and Virginia*. New York: Johnson Reprint Corporation.

Berlin, Isaiah. 1999. *The Roots of Romanticism*. Princeton, NJ: Princeton University Press.

Bickerton, Derek. 1995. *Language and Human Behavior*. Seattle: University of Washington Press.

Bidney, David. 1953. *Theoretical Anthropology*. New York: Columbia University Press.

Binford, Louis. 1962. "Archeology as Anthropology." *American Antiquities* 28.2: 217–225.

Blacking, John. 1992. "Ethnomusicology." In Richard Bauman (ed). *Folklore, Cultural Performances and Popular Entertainment*. New York: Oxford University Press.

Bloomfield, Leonard. 1933. *Language*. New York: Holt.

Boas, Franz. 1991 edition of *Handbook of American Indian Languages*. Published in 1911 in Bulletin 40, Part I, Bureau of American Ethnology (Washington, DC: Government Printing Office. Reprinted in 1966 by the University of Nebraska Press and in paperback in 1991.

Boas, Franz. 1945. *Race, Language and Culture*. New York: Macmillan.

Boas, Franz. 1939. "A Report of the Committee on Research in American Native Languages 1927–1937." *American Council of Learned Societies Bulletin*. no. 29: 105–115.

Botting, Douglas. 1973. *Humboldt and the Cosmos*. New York: Harper and Row Publishers.

Bowman, Isaiah. 1916. *The Andes of Southern Peru: Geographic Reconnaissance Along the Seventy Third Meridian*. New York: Henry Holt and Company.

Brinton, Daniel Garrison. 1885. *The Philosophic Grammar of American Languages, as set forth by Wilhem von Humboldt; with translation of an unpublished memoir by him on the American verb*. Philadelphia.

Bunzi, Matti. 1996. "Franz Boas and the Humboldtian Tradition: From Volksgeist and Nationalcharakter to an Anthropological Concept of Culture." In Stocking, George(ed.) *Volksgeist as Method in Ethnography and the German Anthropological Tradition*. Madison, WI: University of Wisconsin Press.

Burns, Robert "To A Mouse." In George B. Woods, Homer A. Watt, and George K. Anderson. 1941. *The Literature of England: an Anthology and a History*. Chicago: Scott Foresman and Company.

Carpenter, Edmund. 1972. *Eskimo Realities*. New York.

Carter, Ronald. 1982. "Introduction," In Ronald Carter (ed)*Language and Literature: an Introductory Reader in Stylistics*. London: Routledge pp 1–17.

Casson 1981. *Language, Culture and Cognition*. New York: MacMillan Publishing Co. Inc.

Chantell, Glynnis. 2002. *The Oxford Dictionary of Word Histories*. Oxford, UK: Oxford University Press.

Charney, Leo and Vanessa R. Schwartz (eds.) 1995. *Cinema and the Invention of Modern Life*. Berkeley, CA: University of California Press.

Chomsky, Noam. 1972. *Language and Mind*. New York: Harcourt Brace Jovanovich, Inc.

Clifford, James and George Marcus (eds.) 1986. *Writing Culture: The Poetics and Politics of Ethnography*. Berkeley, CA: University of California Press.

Cobb, Matthew. 2015. *Life's Greatest Secret: The Race to Break the Genetic Code*. New York: Basic Books.

Cohen, Jeffrey and Ibrahim Sirkeci. 2011. *Cultures of Migration: the Global Nature of Contemporary Mobility*. Austin TX: University of Texas Press.

Coote, Jeremy and Anthony Shelton (eds). 1992. *Anthropology, Art and Aesthetics*. Oxford: Clarendon Press.

Cushing, Frank. 1979. *Zuñi: Selected Writings of Frank Hamilton Cushing*. University of Nebraska Press.

Custred, Glynn. 1974. "Llameros y comercio interregional." In *Reciprocidad e Intercambioenlos Andes Peruanos*.Giorgio Alberti and Enrique Meyer, (eds.). Peru Problema 12. Instituto de EstudiosPeruanos, Lima, Peru.

Custred, Glynn. 1977a . "Peasant Kinship, Subsistence and Economics in a High Altitude Andean Environment", In *Andean Kinship and Marriage,* Ralph Bolton and Enrique Mayer, (eds.). Special Publication of the American Anthropological Association, No. 7.

Custred, Glynn. 1977b. "Las Punas de los Andes Centrales," 1977. In *Pastores de Puna: Uywamichik Punarunakuna,* Jorge Flores Ochoa, editor. Instituto de Estudios Peruanos, Lima, Peru.

Custred, Glynn. 1979a. "Symbols and Control in a High Altitude Andean Environment." *Anthropos* 1.74.

Custred, Glynn. 1979b. "Inca Concepts of Soul and Spirit." In, Bruce Grindal and Donald M. Warren (eds) *Essays in Humanistic Anthropology: A Festschrift in Honor of David Bidney*. Washington, DC: University Press of America.

Custred, Glynn. 1980a. "The Place of Ritual in Andean Rural Society." In Benjamin Orlove and Glynn Custred (eds). *Land and Power in Latin America.* New York: Holmes and Meir, Publishing, Inc.

Custred, Glynn. 1980b. "Etnolinguistics and the Formulation of Agricultural and Population Censuses." In David Brokensha, Dennis M. Warren and Oswald Werner (eds). *Indigenous Knowledge Systems and Development*. Washington DC: University Press of America.

Custred, Glynn. 1992. "Language Spread and Language Shift in the Central Andes under Spanish Colonial Rule. In eter Nelde (ed). *It's Easier To Mingle If You're Bilingual, Festschrift for Wolfgang Wölck.* Center for Research on Plurilingualism, Brussels. Bonn: Dummler Verlag.

Custred, Glynn. 2011. "The Linguistic Consequences of Boundaries, Borderlands and Frontiers". *Journal of Borderland Studies* 26.3: 265–278.

D'Andrade, Roy. 1995. *The Development of Cognitive Anthropology*. Cambridge, UK: Cambridge University Press.

Daniel, Glyn. 1963. *The Idea of Prehistory*. Cleveland, OH: World Publishing Company.

Darnell, Regna, Michael Hamilton, Michelle Hamilton, Robert Hancock, and Joshua Smith. 2015. *The Franz Boas Papers—Theory, Ethnography, Activism*. Vol.I. University of Nebraska Press.

Diamond, Jared. 2006. *Collapse: How Societies Choose to Fail or Succeed*. New York: Viking Press.

Dilthy, Wilhelm. 1988. *Introduction to the Human Sciences*. Detroit: Wayne State University Press.

Dorson, Richard. 1966. "Introduction," Kurt Ranke. *Folktales of Germany*. Chicago: University of Chicago Press.

Douglas, Mary. 1966. *Purity and Danger: An Analysis of Concepts of Pollution and Taboo.* New York: Praeger.

Douglas, Mary. 1970. *NaturalSymbols: Explorations in Cosmology*. New York: Pantheon Books.

Duranti, Alessandro. 1992. "Oratory" In, Richard Bauman (ed). *Folklore, Cultural Performances and Popular Entertainment*. New York: Oxford University Press.

Duranti, Alessandro. 1997. *Linguistic Anthropology*. Cambridge, UK: Cambridge University Press.

Dutton, Lee (ed). 1999. *Anthropological Resources: A Guide to Archival, Library, and Museum Collections*. New York: Garland Publishing, Inc.

Ellis, John. 1983. *One Fairy Story Too Many: The Brothers Grimm and Their Tales*. Chicago: University of Chicago Press.

Eriksen, Thomas H. and Finn S. Nielsen. 2001. *A History of Anthropology*. London: Pluto Press.

Evans-Pritchard, E. E. 1937. *Witchcraft, Oracles and Magic Among the Azande*. Oxford: Clarendon.

Evans-Pritchard, E. E. 1981. *A History of Anthropological Thought*. New York: Basic Books.

Evans-Pritchard. E. E. 1956. *NuerReligion*. Oxford: Clarendon Press.

Fagan, Brian. 1970. *Introductory Readings in Archeology*. Boston: Little Brown.

Fagan, Brian 1999. *Floods, Famines and the Fate of Civilizations: El Niño and the Fate of Civilizations*. New York: Basic Books.

Ferguson, Niall. 2008. *The Ascent of Money: A Financial History of the World*. New York: Penguin Books.

Firth, Raymond. 1992 "Art and Anthropology." In Jeremy Coote and Anthony Shelton (eds). *Anthropology, Art and Aesthetics*. Oxford: Clarendon Press.

Fortes, Myers and E. E. Evans-Pritchard (eds). *African Political Systems*. Oxford: Clarendon.

Friedrich, Johannes. 1957. *Extinct Languages*. New York: The Philosophical Library.

Frye, Northrop. 1957. *The Anatomy of Criticism*. Princeton, NJ: Princeton University Press.

Geertz, Clifford. 1973. *Interpretation of Cultures*. New York: Basic Books.

Gell, Albert. 1998. *Art and Agency: Towards an Anthropological Theory*. Oxford, UK: Oxford University Press.

Gellner, Ernest. 1983. *Nations and Nationalism*. Ithaca, NY: Cornell University Press.

Gingrich, Andre. 2005. "The German Speaking Countries." In *One Discipline, Four Ways: British, German French and American Anthropology*. The Halle Lectures. Chicago: University of Chicago Press.

Gladwin, Thomas. 1970. *The East is a Big Bird: Navigation and Logic on a Puluwat Atoll.* Cambridge, MA: Harvard University Press.

Glassie, Henry. 1995. *Passing the Time in Ballymenone: Culture and History of an Ulster Community*. Bloomington, IN: Indiana University Press.

Goddard, Ives and William Shipley. 2003. On Linguistics as Evidence. *BriefofAmicus-Curia* in Support of the Plaintiff Appellees in the United States Court of Appeals for the Ninth District. Robson Bonnichsen, et al. v. The United States, et. al. http://friendsofpast.org/pdf/amici/amici-linguistics.pdf.

Goodenough, Ward. 1963. "Cultural Anthropology and Linguistics." Dell Hymes (ed) *Language in Culture and Society*. New York: Harper and Row, Publishers.

Goody, Jack (ed). 1968. *Literacy in Traditional Society*. Cambridge, UK: Cambridge University Press.

Goody, Jack. 1986. *The Logic of Writing and the Organization of Society*. Cambridge, UK: Cambridge University Press.

Goody, Jack. 1987. *The Interface between the Written and the Oral*. Cambridge, UK: Cambridge University Press.

Goody, Jack. 2000. *The Power of the Written Tradition*. Washington, DC: Smithsonian Press.

Goody, Jack and Ian Watt. 1963. *The Consequences of Literacy* In Comparative Studies in History and Society. 5.3: 304–345. Cambridge, UK: Cambridge University Press.

Graburn, Nelson. 2014. "Inuit Native Art: Miracle of the Twentieth Century." The Thomas Garden Barnes Lecture. Canadian Studies Program, University of California, Berkeley.

Gray, Thomas "Elegy Written in a Country Churchyard." In George B. Woods, Homer A. Watt, and George K. Anderson. 1941. *The Literature of England: an Anthology and a History*. Chicago: Scott Foresman and Company.

George B. Woods, Homer A. Watt and George K. Anderson. 56. Chicago: Scott, Foresman and Company.

Green, Jesse. 1979. "Introduction." *Zuni: Selected Writings of Frank Hamilton Cushing*. Lincoln, NE: University of Nebraska Press.

Greenslade, Stanley. 1963. *Cambridge History of the Bible*. vol 1. Cambridge, UK: Cambridge University Press.

Gumperz, John. 1968. "The Speech Community." *International Encyclopedia of the Social Sciences*. 9: 381–386.

Gumperz, John, 1982a. *Discourse Strategies*. New York: Cambridge University Press.

Gumperz, John, 1982b. *Language and Social Identity*. New York: Cambridge University Press.

Hann, Chris and Keth Hart. *Economic Anthropology: History, Ethnography and Critique*. Polity Books.

Heath, Shirley and R. Lapade. 1982. "Castellian Colonization and Indian Languages: The Case of Quechua and Aymara." In Robert Cooper (ed). *Language Spread*. Bloomington, IN: Indiana University Press.

Heckewelder, John, 1972 (Reprint). *History of the Manners and Customs of the Indian Nations That Once Inhabited Pennsylvania and the Neighboring States*. New York: Arno Press.

Heider, Karl G. 2006. *Ethnographic Film* (revised edition). Austin: University of Texas Press.

Hemming, John. 1970. *The Conquest of the Incas*. New York: Harcourt, Brime, and Janovich.

Hinsley, Curtis. 1983. "Ethnographic Charisma and Scientific Routine: Cushing and Fewkes in the American Southwest, 1879–1893." In George Stocking (ed) *Observers Observed: Essays on Ethnographic Fieldwork*, pp. 53–69.

Hodge, H. A. 1969 [1944]. *Wilhelm Dilthey: An Introduction*. New York: Howard Fertig.

Hodge, H. A. 1952. *The Philosophy of Wilhelm Dilthey*. Westport, CT: Greenwood Press, Publishers.

Hoebel, Adamson, Richard L. Currier and S. Kaiser. 1982. *Crisis in Anthropology: View from Spring Hill 1980*. New York: Garland Publishing.

Holmes, R. 2008. *The Age of Wonder: How the Romantic Generation Discovered the Beauty and Terror of Science*. London: Harper Press.

Horton, Robin. 1970. "African Thought and Western Science." In Max Marwick (ed) *Witchcraft and Sorcery*. Hamondsworth: Penguin ooks, Inc. 342–368.

Huizinga, Johan. 1955. *Homo ludens: a Study of the Play-element in Culture*. Chicago: University of Chicago Press.

Hull, David. 1988. *Science as a Process*. Chicago: University of Chicago Press.

Humboldt, Alexander. 1849. *Cosmos: A Sketch of a Physical Description of the Universe*. Vol. 1 Covant Gardens: Henry G. Bohn.

Hymes, Dell. 1964. *Language in Culture and Society*. New York: Harper and Row Publishers.

Hymes, Dell. 1968a. "The Ethnography of Speaking." In Joshua Fishman (ed). *Readings in the Sociology of Language*. The Hague: Mouton.

Hymes, Dell. 1968b. "Linguistics: The Field." *The International Encyclopedia of the Social Sciences*. 9: 351–371.

Hymes, Dell. 1983. *Essays in the History of Linguistic Anthropology*. Amsterdam/Philadelphia: John Benjamin's Publishing Company.

Johanson, Donald and Kate Wong. 2010. *Lucy's Legacy: The Quest for Human Origins*. New York: Random House, Inc.

Keil, Charles. 1966. *Urban Blues*. Chicago: University of Chicago Press.

Keil, Charles. 1979. *Tiv Song*. Chicago: University of Chicago Press.

Keil, Charles. 2002. *Bright Balkan Morning: Romani Lives and the Power of Music*. Middletown, CT: Wesleyan University Press.

Kent, Johnathan. 1982. *The Domestication and Exploitation of the South American Camelids*. PhD diss., Washington University.

Kluckhohn, Clyde and Oralf Prufer. 1959. "Influences During the Formative Years." In Goldschmidt, Walter (ed). *The Anthropology of Franz Boas*. Memoire no. 89 of the American Anthropological Association Volume 61, No. 5, Part 2, 4–28.

Kroeber, A. L. 1943. *Franz Boas*. Menasha, WI: American Anthropological Assciation.

Kroeber, A. L. and Clyde Kluckhohn. 1952. *Culture: a Critical Review of Concepts andDefinitions*. New York: Vantage Books.

Kryza, Frank. 2006. *The Race for Timbuktu: In Search of the City of Gold*. New York. Ecco.

Kubler, George. 1946. "The Quechau in the Colonial World." In Julian Steward (ed). The Handbook of South American Indians. Vol. 2. *The Andean Civilizations*. Washington, DC: Smithsonian Institution.

Kuhn, Thomas S. Kuhn. 1970. The Structure of Scientific Revolutions. Chicago: University of Chicago Press.

Kuper, Adam. 1973. *Anthropology and Anthropologists: The British School 1922–1972*. London: Allen Lane.

Kuper, Adam. 1988. *The Invention of Prehistory*. London: Routledge.

Kuznar, Lawrence. 1995. *Awaticmark: The Ethnoarcheology of an Andean Herding Community*. New York: Harcourt Brace.

Kuznar, Lawrence. 1997 First Edition. *Reclaiming a Scientific Anthropology*. Walnut Creek, CA: Altamira Press, Second Edition in 2008, Lanham, MD: Altamira Press.

Kuznar, Lawrence (ed). 2001. *Ethnoarcheology of Andean South America: Contributions to Archeological Method and Theory*. International Monographs in Prehistory.

Lanning, Edward. 1967. *Peru Before the Incas*. Englewood Cliffs, NJ: Printice Hall.

Lesser, Alexander. 1968. "Boas, Franz." *International Encyclopedia of the Social Sciences* 2: 99-110.

Layton, Robert. 1981. *The Anthropology of Art*. London: Granada Publishing.

Lenneberg, Eric. 1967. *The Biological Foundations of Language*. New York: Wiley.

Levy-Strauss, Claude. 1962–1967. *Structural Anthropology*. Vols 1 and 2. New York: Basic Books.

Levy-Strauss, Claude. 1966. *The Savage Mind*. Chicago: The University of Chicago Press.

Levy-Strauss, Claude. 1969. *Elementary Structures of Kinship*. Boston: Beacon Press.

Levy-Strauss, Claude. 1978. *Myth and Meaning*. London: Routledge and Kegan Paul.

Lewis, Herbert S. 2014. *In Defense of Anthropology: An Investigation of the Critique of Anthropology*. New Brunswick, NJ: Transaction Publishers.

Lindberg, David. 1992. *The Beginnings of Science: the European Scientific Tradition in Philosophy, Religion and Institutional Context 600 BC to 1450*. Chicago: University of Chicago Press.

Lecoq, P. 1987. "Caravens de lamas, Sel et Echangedansune Communauté de Potosi, en Bolivie. *Bulletin de l'Institut Francasisd" Etudes Andines*. 16.3–4: 1–38.

Lecoq, P. 1988. UnaRuta 'de la Sal' en el Sud Boliviano: Informedel Viaje de Truque Annual de una caravan de Llamas. *Revista del Museo Nacional de Etnografía y Folklore*. 1–2: 163–216. La Paz.

Lowie, Robert. 1937. *The History of Ethnological Theory*. New York: Holt, Rinehart and Winston.

Lowie, Robert. 1956. *Reminiscence of Anthropological Currents America Half a Century Ago*. American Anthropological Association. New Series 58.6: 995–1016.

Makkreel, Rudolp and Sebastian Luft. 2010. "Dilthey and the Neo-Kantians: The Dispute over the Status of the Cultural Sciences," In Dean Moyar (ed). *The Rutledge Companion to Nineteenth Century Philosophy*. London: Rutledge. 435–466.

Malinowski, Bronislaw. 1944. *A Scientific Theory of Culture: and Other Essays*. Chapel Hill: University of North Carolina Press.

Marwick, Max (ed). 1970. *Witchcraft and Sorcery*. Harmondsworth, Middlesex, England: Penguin Books, Ltd.

Medawer, Peter. 1967. *The Art of the Soluble*. London: Methuen, and Co, Ltd.

Merriam, Alan. 1969. *The Anthropology of Music*. Evanston, IL: Northwestern University Press.

Metraux, Rhoda. 1968. "Malinowski, Bronislaw." *International Encyclopedia of the Social Sciences*. 9: 541-549.

Miller, George. 1979. *An Introduction to Ethnoarchiology of the Andean Camelids*. PhD diss., University of California Berkeley.

Molina Rivera, R. 1987. In O. Harris, B. Larson and E. Tandeter (eds). *La Participación Indígenaen los Mercados Surandinos*. La Paz: Centro de Estudios de la Realidad Económico y Social pp. 603–636.

Morris, Craig. 1986. "Storage, Supply and Redistribution in the Economy of the Inka State," In John V. Murra, Nathan Wachtel and Jacques Revel (eds). *Anthropological History of Andean Polities*. Cambridge, UK: Cambridge University Press 59–68.

Munro, Thomas. 1967. *The Arts and Their Interrelations*. Cleveland, OH: The Press of Western Reserve University.

Murra, John V. 1960. "Rite and Crop in the Inca State." In Stanley Diamond (ed) *Culture in History*. New York:

Murra, John V. 1962. "La Función del Tejído en Varios Contextos Sociales del Estado del Inca." *Actas y Trabajos*. Lima, Peru: Secundo Congreso Nacional del Peru. 1958: 215–240.

Murra, John V. 1972. *"El Control Vertical" de un Máximo Pisos Ecológicos en la Economía de las Sociedades Andinas."* Huanuco, Peru: Universidad Hermilo Valdizán.

Murra, John. 1986. "The Expansion of the Inka State: Armies, Wars and Rebellions." In John V. Murra, Nathan Wachtel, and Jacques Revel (eds). *Anthropological History of Andean Polities*. Cambridge, UK: Cambridge University Press 49–58.

Nagel, Ernest. 1961. *The Structure of Science: Problems in the Logic of Scientific Explanation*. London: Routledge and Kegan Paul.

Netting, Robert M. 1982. "The Ecological Perspective: Holism and Scholasticism in Anthropology." E. A. Hoebel, R. Currier, S. Kaiser (eds). *Crisis in Anthropology*. New York: Garland.

Netting, Robert M. 1986. *Cultural Ecology*. Second Edition. Long Grove, IL: Waveland Press.

Nielsen, Axel E. 2001. "Ethnoarchaeological Perspectives on Caravan Trade in the South Central Andes." In Lawrence Kuznar (ed). *Ethnoarchaeology of Andean South America: Contributions to Archaeological Method and Theory*. International monographs in prehistory.

Olson, David R. 1994. *The World On Paper: The Conceptual and Cognitive Implications of Writing and Reading*. Cambridge, UK: Cambridge University Press.

Osborne, Harold. 1970. *The Art of Appreciation*. Oxford, UK: Oxford University Press.

Outwaite, William. 1975. *Understanding Social Life*. London: George Allen and Unwin.

Parker, Geoffrey. 2013. *Global Crisis: War, Climate Change and Catastrophe in the Seventeenth Century*. New Haven, CT: Yale University Press.

Parker, Geoffrey. 1980. *Europe in Crisis: 1598–1648*. Ithaca, NY: Cornell University Press.

Parkin, Robert. 2005. "The French Speaking Countries" In, *One Discipline, Four Ways: British, German French and American Anthropology*. The Halle Lectures. Chicago: University of Chicago Press.

Porras Barranchea, Rául. 1963. *Fuentes Historicos del Peru*. Lima, Peru: Universidad Nacional Mayor de San Marcos.

Prescott, William H. *The History of the Conquest of Peru*. Reprint 2001 by Cosimo.

Ready, Elspeth. 2015 (December 1). "Food Security and Economic Interdependence in the Canadian Arctic." Canadian Studies Colloquium, University of California, Berkeley.

Renfrew, Colin. 2003. *Figuring It Out. Where Do We Come From? The Parallel Visions Artists and Archeologists*. Cambridge, UK: Cambridge University Press.

Renfrew, Colin and Paul Bahn. 1991. *Archaeology: Theories, Methods and Practice*. New York: Thames Hudson Inc.

Rickert, Heinrich. 1986. *The Limits of Concept Formation in Natural Science*. Cambridge, UK: Cambridge University Press.

Rowe, John. 1946. "Inca Culture at the Time of the Spanish Conquest. In, Julian Steward (ed). The handbook of South American Indians. Vol. 2. *The Andean Civilizations*. Washington, DC: Smithsonian Institution.

Rowe, John. 1960. "The Origins of Creator Worship Among the Incas." In Stanley Diamond (ed) *Culture in History*. New York:

Salmon, Frank. 2001. "How an Andean Writing 'Without Words' Works." *Current Anthropology*.

Salmon, Frank and Mercedes Niño-Murcia. 2011. *The Lettered Mountain: A Peruvian Village's Way of Writing*. Durham, NC: Duke University Press.

Schaff, Adam. 1973. *Language and Cognition*. New York: McGraw-Hill.

Scharff, Robert C. 2010. "Comte's Positivist Dream, Our Positivist Burden." In Dean Moyer (ed). *Rutledge Companion to Nineteenth Century Philosophy*. London: Rutledge.

Sapir, Edward. 1921. *Language: An Introduction to the Study of Speech*. San Diego, CA: Harcourt Brace & Co.

Sapir, Edward. 1929. "The Status of Linguistics as a Science." Language 5: 207–214.

Scribner, Silvia and Michael Cole. 1981. *The Psychology of Literacy*. Cambridge, MA: Harvard University Press.

Shelton, Anthony. 1992. "Predicates and Aesthetic Judgments: Ontology and Value in Huichol Material Representations." In Jeremy Coote and Anthony Shelton (eds). *Anthropology, Art and Aesthetics*. Oxford: Clarendon Press. 209–244.

Shipman, Pat. 2015. *Invasion: How Humans and Their Dogs Drove Neanderthal Man to Extinction*. Belknap Press.

Schneider, David. 1972. "What is Kinship All About?" In P. Reinig (ed). *Kinship Studies in the Morgan Memorial Year*. Washington, DC: Anthropological Society of Washington.

Schneider, David. 1984. *A Critique of the Study of Kinship*. Ann Arbor: University of Michigan Press.

Shreeve, Jamie. 2015 (Sept). "Homo Naledi: A New Discovery, Hidden in a Hard-to-Access Cave." *National Geographic*.

Schrödinger, Erwin. 2014. *Nature and the Greeks and Science and Humanism*. Cambridge, UK: Cambridge University Press.

Silverman, Sydel. 2005. "The United States." In Chris Hann (ed). *One Discipline, Four Ways: British, German French and American Anthropology*. Chicago: Universityof Chicago Press.

Simic, Andrei and Glynn Custred. 2003. On Oral Tradition as Evidence. *Brief of Amicus Curia* in Support of the Plaintiff Appellees in the United States Court of Appeals for the Ninth District. Robson Bonnichsen, et al. v. The United States, et. al. http://friendsofpast.org/pdf/amici/amici-oral_traditions.pdf.

Smith, Anthony D. 1986. *The Ethnic Origin of Nations*. Oxford: Basil Blackwell.

Spier, Leslie. 1959. "Some Central Elements in the Legacy," In Goldschmidt, Walter (ed). *The Anthropology of Franz Boas*. Memoire no. 89 of the American Anthropological Association Volume 61, No. 5, Part 2, 146–155.

Spradley, James P. and George E. McDonough (eds). *Anthropology Through Literature*. Boston: Little Brown.

Spradley, James P. 1980. *Participant Observation*. New York: Holt, Rinehart and Winston.

Stemmer, Brigitte and Harry A. Whitaker (eds). 1988. *Handbook of Neurolinguistics*. Academic Press.

Steward, Julian. 1973. *Alfred Kroeber*. New York: Columbia University Press).

Stocking, George. 1965. "From Physics to Ethnology: Franz Boas's Arctic Expedition as a Problem in the Historiography of the Behavioral Science." *Journal of the History of the Behavioral Sciences* 1.1:53–66.

Stocking, George. 1968. *Race, Culture and Evolution: Essays in the History of Anthropology*. New York: Collier-Macmillan.

Stocking, George (ed).1983. *Observers Observed: Essays on Ethnographic Fieldwork* Madison: University of Wisconsin Press.

Stocking, George. 1987. *Victorian Anthropology*. New York: The Free Press.

Stocking, George (ed). 1996. *Volksgeist as Method in Ethnography and the German Anthropological Tradition*. Madison, WI: the University of Wisconsin Press.

Suddendorf, Thomas. *The Gap: the Science that Separates Us from the Other Animals*. New York: Basic Books.

Swinton, G. 1978. "Touch and the Real: Contemporary Inuit Aesthetics." In M. Greenhalghand V. Megaw (eds). *Art in Society: Studies in Style, Culture and Aesthetics*. London: Duckworth.

Tax, Sol (ed). 1962. *Anthropology Today: Selections(Structural anthro)*. Chicago: University of Chicago Press.

Tambiah, Stanley. 1990. *Magic, Science, Religion and the Scope of Rationality*. Cambridge, UK: Cambridge University Press.

Tannen, Deborah (ed). 1981. "Introduction" *Analyzing Discourse*. Washington, DC: Georgetown University Press.

Tetens, Holme. 2013. *Wissenschaftstheorie: Eine Einführung*. Munich, Germany: Verlag C. H. Beck.

Tosi, Joseph. 1960. *Zonas de Vida Natural en el Peru*. Lima, Peru: Instituto Interamericanode Ciencias Agricolas de la OEA.

Truzzi, Marcello (ed) 1974. *Verstehen: Subjective Understanding in the Social Sciences*. Addison-Wesley Publishing Company.

Urton, Gary. 2003. *Signs of the Khipu*. Binary Coding in the Andean Knotted String Records. Austin: University of Texas Press.

Von Hayek, Friedrich. 1955. *The Counter Revolution of Science: Studies on the Abuse of Reason*. London: The Free Press of Glencoe.

Wade, Nicholas. 2010 (12/9). "Anthropology a Science? Statement Deepens a Rift." *New York Times*.

Wachtel, Nathan. 1973. *Sociedad e Ideología: Ensayos de Historia y Anthropología Andinas*. Lima, Peru: Instituto de Estudios Peruanos.

Wallace, Anthony F. C. 1965. "Introduction" to *James Mooney: The Ghost Dance Religion and the Sioux Outbreak of 1890*. Chicago: University of Chicago Press.

Wallace, Anthony F. C. 1966. *Religion: An Anthropological View.* New York: Random House.

Walter, Chip. 2015 (January)."First Artists." *National Geographic*. 32–57.

Weakland, John H. 1975. "Feature Films as Cultural Documents." In Paul Hockings (ed). *Principles of Visual Anthropology*. The Hague: Mouton.

Weber, Max. 1972 *Max Weber: The Interpretation of Social Reality*. New York: Schnocken Press.

Weber, David J. and Jane M. Rausch (eds). 1994. *Where Cultures Meet: Frontiers in Latin American History*. Jaguar Books in Latin American History. Wilmington, DE: A Scholarly Resources, Inc. Imprint.

Weinreich, Uriel. 1968. "Is a Structural Dialectology Possible?" In Joshua Fishman (ed). *Readings in the Sociology of Language*. The Hague: Mouton.

West, T. 1981. "Llama Caravans of the Andes." *Natural History* 90.12: 62–73.

Willems, Emilio. 1994. David Weber and Jane M. Rausch (eds). 1994. *Where Cultures Meet: Frontiers in Latin American History*. Jaguar Books in Latin American History. Wilmington, DE: A Scholarly Resources, Inc. Imprint.

Williams, Glyn 2013. *Naturalists at Sea: Scientific Travelers from Dampier to Darwin*. New Haven, CT: Yale University Press.

Wilson, E. O. 1998. *Consilience: The Unity ofKnowledge*. New York: Alfred Knopf.

Wolczek, Frank. 2015. Interview "Die Welt Ist Ein Kunstwerk," *Der Spiegel* 33.

Wolf, Eric. 1974. *Anthropology*. New York: Norton.

Wulf, Andrea. 2015. *The Invention of Nature: Alexander von Humboldt's New World*. New York: Alfred A. Knopf.

Index

About the Author

Glynn Custred is a cultural anthropologist who has worked in the Central Andean highlands and in Europe. He has taught general anthropology (the four fields approach), cultural anthropology, folklore, and linguistics at California State University–East Bay where he is now a professor emeritus.